GENE CLONING
AND DNA ANALYSIS
An Introduction

GENE CLONING AND DNA ANALYSIS

An Introduction

T.A. BROWN
Department of Biomolecular Sciences
UMIST
P.O. Box 88
Manchester
M60 1QD
UK

Fourth Edition

**Blackwell
Science**

© 2001 by
Blackwell Science Ltd
Editorial Offices:
Osney Mead, Oxford OX2 0EL
25 John Street, London WC1N 2BS
23 Ainslie Place, Edinburgh EH3 6AJ
350 Main Street, Malden
  MA 02148 5018, USA
54 University Street, Carlton
  Victoria 3053, Australia
10, rue Casimir Delavigne
  75006 Paris, France

Other Editorial Offices:

Blackwell Wissenschafts-Verlag GmbH
Kurfürstendamm 57
10707 Berlin, Germany

Blackwell Science KK
MG Kodenmacho Building
7–10 Kodenmacho Nihombashi
Chuo-ku, Tokyo 104, Japan

Iowa State University Press
A Blackwell Science Company
2121 S. State Avenue
Ames, Iowa 50014-8300, USA

First, second and third editions
published by Chapman & Hall 1986,
1990, 1995
Reprinted 1998 by
Stanley Thomes (Publishers) Ltd
Fourth edition published by Blackwell
Science Ltd 2001

Set in 10/13.5 pt Times
by Best-set Typesetter Ltd., Hong Kong
Printed and bound in Great Britain by
Ashford Colour Press Ltd, Gosport

The right of the Author to be
identified as the Author of this Work
has been asserted in accordance with
the Copyright, Designs and Patents
Act 1988.

A catalogue record for this title
is available from the British Library

ISBN 0-632-05901-X

Library of Congress
Cataloging-in-Publication Data

Brown, T.A. (Terence A.)
    Gene cloning and DNA analysis /
Terry A. Brown. – 4th ed.
      p. cm.
    Prev. ed. published with title:
Gene cloning.
    Includes bibliographical
references and index.
    ISBN 0-632-05901-X (alk. paper)
    1. Molecular cloning.
2. Nucleotide sequence. 3. DNA –
Analysis. I. Brown, T.A. (Terence
A.). Gene cloning. II. Title.

QH442.2 .B76 2001
572.8′633 – dc21   2001025352

DISTRIBUTORS

Marston Book Services Ltd
PO Box 269
Abingdon
Oxon OX14 4YN
(Orders: Tel: 01235 465500
              Fax: 01235 465555)

USA
Blackwell Science, Inc.
Commerce Place
350 Main Street
Malden, MA 02148 5018
(Orders: Tel: 800 759 6102
                    781 388 8250
              Fax: 781 388 8255)

Canada
Login Brothers Book Company
324 Saulteaux Crescent
Winnipeg, Manitoba R3J 3T2
(Orders: Tel: 204 837-3987
              Fax: 204 837-3116)

Australia
Blackwell Science Pty Ltd
54 University Street
Carlton, Victoria 3053
(Orders: Tel: 03 9347 0300
              Fax: 03 9347 5001)

For further information on
Blackwell Science, visit our website:
www.blackwell-science.com

Contents

Part 3: The Applications of Gene Cloning and DNA Analysis in Biotechnology, 269

Preface to the Fourth Edition

Fifteen years ago, when the first edition of this book was published, gene cloning was a relatively new technique that provided the basis of virtually all studies of DNA. Now non-cloning approaches, notably the polymerase chain reaction (PCR), are equally important. To reflect these changes, the fourth edition has a new title. This does not mean that the philosophy of the book has changed. It is still an introductory text that begins at the beginning and does not assume the reader has any prior knowledge of the techniques used to study genes and genomes. The new title simply means that the breadth of the book is greater than it needed to be in 1986.

This edition has two entirely new chapters and some organizational changes. The first of the new chapters is about the methods used to sequence genomes and to understand the sequence after it has been obtained. This is, of course, much more than simply sequencing on a grand scale and involves strategies for assembling contiguous sequences, methods for identifying genes in a genome sequence, and techniques for studying the transcriptome and proteome. The second new chapter is the final one, on the applications of gene cloning and DNA analysis in forensic science. This is a popular topic with students and one that provides an excellent illustration of the applications of DNA analysis in the real world.

The organizational changes see PCR moved to an earlier position so that its applications in research and biotechnology can be dealt with adequately in Parts 2 and 3. I have also moved a few other sections around to try to produce a more logical flow of information. As usual, there are updates and the few errors have been corrected.

For some time I was unsure if there would be a fourth edition, and I would therefore like to thank Nigel Balmforth of Blackwell Science for ensuring that the book survived the takeover of the previous publishers. I must also thank my colleagues Lubomira Stateva and Keri Brown for allowing me to make use of their teaching materials as cribs for some of the new information in Part 3. At the end of the Preface I usually say how much hard work the new edition has been, but to be honest doing this one was very enjoyable.

T.A. Brown
Department of Biomolecular Sciences
UMIST
Manchester

Preface to the Third Edition

The third edition of this book is longer and more detailed than either of its predecessors, but it retains the same basic philosophy. It is still unashamedly introductory and is still aimed at undergraduates and other individuals who have no previous experience of experiments with DNA. The last few years have seen a proliferation of cloning manuals and other 'hands-on' texts for gene cloners, but few of these address the needs of a student encountering the subject for the first time. I hope that this new edition will continue to help these newcomers get started.

I have made revisions and updates at many places throughout the book, but major changes do not occur until the last few chapters. A major weakness of the second edition was the poor coverage that I had given to the polymerase chain reaction, my excuse being that the sudden rise of the technique occurred just as I had completed the manuscript. The third edition attempts to correct this failing with a new chapter devoted entirely to PCR. I appreciate that PCR is not really a component of gene cloning, so the title of the book is no longer appropriate, but *Gene Cloning, the Polymerase Chain Reaction, and Related Techniques: An Introduction* seemed a bit long-winded. The 'gene cloning' of the title is now something of a generic term, and I apologize to any purists among the readership.

The other major rewrites are in Part 3, where again I have extended the coverage, this time by bringing in a new chapter on the applications of gene cloning in agriculture, or to be precise, in plant genetic engineering. Here, and at a few other places in the book, I have attempted in a rather hesitant (and I fear inadequate) fashion to discuss some of the broader issues that arise from our ability to clone genes. These issues are now so prominent in the public perception that they must be addressed by all students of genetic engineering. So the real title of the book is *Gene Cloning, the Polymerase Chain Reaction, Related Techniques and Some of the Implications: An Introduction.*

As with the first and second editions, I would not have got very far with this book if my wife Keri had not been prepared to put up with several months of lonely evenings as I struggled with the word processor. Once again her unwavering encouragement has been the most important factor in completion of the task.

T.A. Brown
Manchester

Preface to the Second Edition

It was only when I started writing the second edition to this book that I fully appreciated how far gene cloning has progressed since 1986. Being caught up in the day-to-day excitement of biological research it is sometimes difficult to stand back and take a considered view of everything that is going on. The pace with which new techniques have been developed and applied to recombinant DNA research is quite remarkable. Procedures which in 1986 were new and innovative are now *de rigueur* for any self-respecting research laboratory and many of the standard techniques have found their way into undergraduate practical classes. Students are now faced with a vast array of different procedures for cloning genes and an even more diverse set of techniques for studying them once they have been cloned.

In revising this book I have tried to keep rigidly to a self-imposed rule that I would not make the second edition any more advanced than the first. There are any number of advanced texts for students or research workers who need detailed information on individual techniques and approaches. In contrast, there is still a surprising paucity of really introductory texts on gene cloning. The first edition was unashamedly introductory and I hope that the second edition will be also.

Nevertheless, changes were needed and on the whole the second edition contains more information. I have resisted the temptation to make many additions to Part 1, where the fundamentals of gene cloning are covered. A few new vectors are described, especially for cloning in eukaryotes, but on the whole the first seven chapters are very much as they were in the first edition. Part 2 has been redefined so it now concentrates more fully on techniques for studying cloned genes, in particular with a description of methods for analysing gene regulation. Recombinant DNA techniques in general have become more numerous since 1986 and an undergraduate is now expected to have a broader appreciation of how cloned genes are studied. In Part 3 the main theme is still biotechnology, but the tremendous advances in this area have required more extensive rewriting. The use of eukaryotes for synthesis of recombinant protein is now standard procedure, and we have seen the first great contributions of gene cloning to the study of human disease. The applications of gene cloning really make up a different book to this one, but nonetheless in Part 3 I have tried to give a flavour of what is going on.

A number of people have been kind enough to comment on the first edition and make suggestions for this revision. Don Grierson and Paul Sims again provided important and sensible advice. I must also thank Stephen Oliver and Richard Walmsley for their comments on specific parts of the book. Once again my wife's patience and encouragement has been a major factor in getting a second edition done at all. Finally I would like to thank all the students who have used the first edition for the mainly nice things they have said about it.

T.A. Brown
Manchester

Preface to the First Edition

This book is intended to introduce gene cloning and recombinant DNA technology to undergraduates who have no previous experience of the subject. As such, it assumes very little background knowledge on the part of the reader – just the fundamental details of DNA and genes that would be expected of an average sixth-former capable of a university entrance grade at A-level biology. I have tried to explain all the important concepts from first principles, to define all unfamiliar terms either in the text or in the glossary, to avoid the less helpful jargon words, and to reinforce the text with as many figures as are commensurate with a book of reasonable price.

Although aimed specifically at first- and second-year undergraduates in biochemistry and related degree courses, I hope that this book will also prove useful to some experienced researchers. I have been struck over the last few years by the number of biologists, expert in other aspects of the science, who have realized that gene cloning may have a role in their own research projects. Possibly this text can act as a painless introduction to the complexities of recombinant DNA technology for those of my colleagues wishing to branch out into this new discipline.

I would like to make it clear that this book is not intended as competition for the two excellent gene cloning texts already on the market. I have considerable regard for the books by Drs Old and Primrose and by Professor Glover, but believe that both texts are aimed primarily at advanced undergraduates who have had some previous exposure to the subject. It is this 'previous exposure' that I aim to provide. My greatest satisfaction will come if this book is accepted as a primer for Old and Primrose or for Glover.

I underestimated the effort needed to produce such a book and must thank several people for their help. The publishers provided the initial push to get the project under way. I am indebted to Don Grierson at Nottingham University and Paul Sims at UMIST for reading the text and suggesting improvements; all errors and naïveties are, however, mine. Finally, my wife Keri typed most of the manuscript and came to my rescue on several occasions with the right word or turn of phrase. This would never have been finished without her encouragement.

T.A. Brown
Manchester

PART 1
THE BASIC PRINCIPLES OF GENE CLONING AND DNA ANALYSIS

Chapter 1   # Why Gene Cloning and DNA Analysis are Important

In the middle of the nineteenth century, Gregor Mendel formulated a set of rules to explain the inheritance of biological characteristics. The basic assumption of these rules is that each heritable property of an organism is controlled by a factor, called a **gene**, that is a physical particle present somewhere in the cell. The rediscovery of Mendel's laws in 1900 marks the birth of **genetics**, the science aimed at understanding what these genes are and exactly how they work.

1.1   ## The early development of genetics

For the first 30 years of its life this new science grew at an astonishing rate. The idea that genes reside on **chromosomes** was proposed by W. Sutton in 1903, and received experimental backing from T.H. Morgan in 1910. Morgan and his colleagues then developed the techniques for **gene mapping**, and by 1922 had produced a comprehensive analysis of the relative positions of over 2000 genes on the four chromosomes of the fruit fly, *Drosophila melanogaster*.

Despite the brilliance of these classical genetic studies, there was no real understanding of the molecular nature of the gene until the 1940s. Indeed, it was not until the experiments of Avery, MacLeod and McCarty in 1944, and of Hershey and Chase in 1952, that anyone believed deoxyribonucleic acid (DNA) to be the genetic material: up until then it was widely thought that genes were made of protein. The discovery of the role of DNA was a tremendous stimulus to genetic research, and many famous biologists (Delbrück, Chargaff, Crick and Monod were among the most influential) contributed to the second great age of genetics. In the 14 years between 1952 and 1966 the structure of DNA was elucidated, the genetic code cracked, and the processes of transcription and translation described.

1.2

The advent of gene cloning and the polymerase chain reaction

These years of activity and discovery were followed by a lull, a period of anticlimax when it seemed to some molecular biologists (as the new generation of geneticists styled themselves) that there was little of fundamental importance that was not understood. In truth there was a frustration that the experimental techniques of the late 1960s were not sophisticated enough to allow the gene to be studied in any greater detail.

Then in the years 1971–1973 genetic research was thrown back into gear by what at the time was described as a revolution in experimental biology. A whole new methodology was developed, enabling previously impossible experiments to be planned and carried out, if not with ease, then at least with success. These methods, referred to as **recombinant DNA technology** or **genetic engineering**, and having at their core the process of **gene cloning**, sparked another great age of genetics. They led to rapid and efficient DNA sequencing techniques that enabled the structures of individual genes to be determined, reaching a culmination in the 1990s with the massive genome sequencing projects, including the human project which was completed in 2000. They led to procedures for studying the regulation of individual genes, which have allowed molecular biologists to understand how aberrations in gene regulation can result in human diseases such as cancer. The techniques spawned modern **biotechnology**, which puts genes to work in production of proteins and other compounds needed in medicine and industrial processes.

During the 1980s, when the excitement engendered by the gene cloning revolution was at its height, it hardly seemed possible that another, equally novel and equally revolutionary process was just around the corner. According to DNA folklore, Kary Mullis invented the **polymerase chain reaction (PCR)** during a drive along the coast of California one evening in 1985. His brainwave was an exquisitely simple technique that acts as a perfect complement to gene cloning. PCR has made easier many of the techniques that were possible but difficult to carry out when gene cloning was used on its own. It has extended the range of DNA analysis and led to molecular biology finding new applications in areas of endeavour outside of its traditional range of medicine, agriculture and biotechnology. Molecular ecology, biomolecular archaeology and DNA forensics are just three of the new disciplines that have become possible as a direct consequence of the invention of PCR, and molecular biologists are devising new ways of using DNA to ask questions about human evolution and the impact of environmental change on the biosphere. Thirty years after the gene cloning revolution we are still riding the rollercoaster and there is no end to the excitement in sight.

1.3

What is gene cloning?

The basic steps in a gene cloning experiment are as follows (Figure 1.1).

(1) A fragment of DNA, containing the gene to be cloned, is inserted into a circular DNA molecule called a **vector**, to produce a **chimera** or **recombinant DNA molecule**.

Figure 1.1 The basic steps in gene cloning.

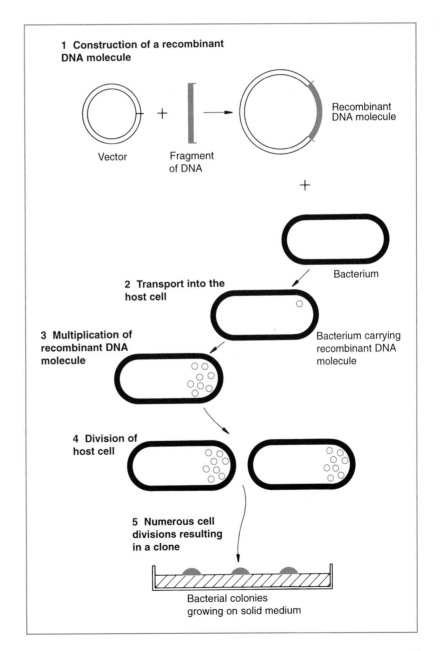

1 **Construction of a recombinant DNA molecule**

Vector   Fragment of DNA   Recombinant DNA molecule

Bacterium

2 **Transport into the host cell**

3 **Multiplication of recombinant DNA molecule**

Bacterium carrying recombinant DNA molecule

4 **Division of host cell**

5 **Numerous cell divisions resulting in a clone**

Bacterial colonies growing on solid medium

(2) The vector acts as a **vehicle** that transports the gene into a host cell, which is usually a bacterium, although other types of living cell can be used.

(3) Within the host cell the vector multiplies, producing numerous identical copies not only of itself but also of the gene that it carries.

(4) When the host cell divides, copies of the recombinant DNA molecule are passed to the progeny and further vector replication takes place.

(5) After a large number of cell divisions, a colony, or **clone**, of identical host cells is produced. Each cell in the clone contains one or more copies of the recombinant DNA molecule; the gene carried by the recombinant molecule is now said to be cloned.

1.4   What is PCR?

The polymerase chain reaction is very different from gene cloning. Rather than a series of manipulations involving living cells, PCR is carried out in a single test tube simply by mixing DNA with a set of reagents and placing the tube in a thermal cycler, a piece of equipment that enables the mixture to be incubated at a series of temperatures that are varied in a preprogrammed manner. The basic steps in a PCR experiment are as follows (Figure 1.2):

(1) The mixture is heated to 94°C, at which temperature the hydrogen bonds that hold together the two strands of the double-stranded DNA molecule are broken, causing the molecule to **denature**.

(2) The mixture is cooled down to 50–60°C. The two strands of each molecule could join back together at this temperature, but most do not because the mixture contains a large excess of short DNA molecules, called **oligonucleotides** or **primers**, which **anneal** to the DNA molecules at specific positions.

(3) The temperature is raised to 74°C. This is the optimum working temperature for the ***Taq* DNA polymerase** that is present in the mixture. We will learn more about **DNA polymerases** on p. 56. All we need to understand at this stage is that the *Taq* DNA polymerase attaches to one end of each primer and synthesizes new strands of DNA, complementary to the **template** DNA molecules, during this step of the PCR. Now we have four stands of DNA instead of the two that there were to start with.

(4) The temperature is increased back to 94°C. The double-stranded DNA molecules, each of which consists of one strand of the original molecule and one new strand of DNA, denature into single strands. This begins a second cycle of denaturation–annealing–synthesis, at the end of which there are eight DNA strands. By repeating the cycle 25 times the double-stranded molecule that we began with is converted into over 50

Figure 1.2 The basic steps in the polymerase chain reaction.

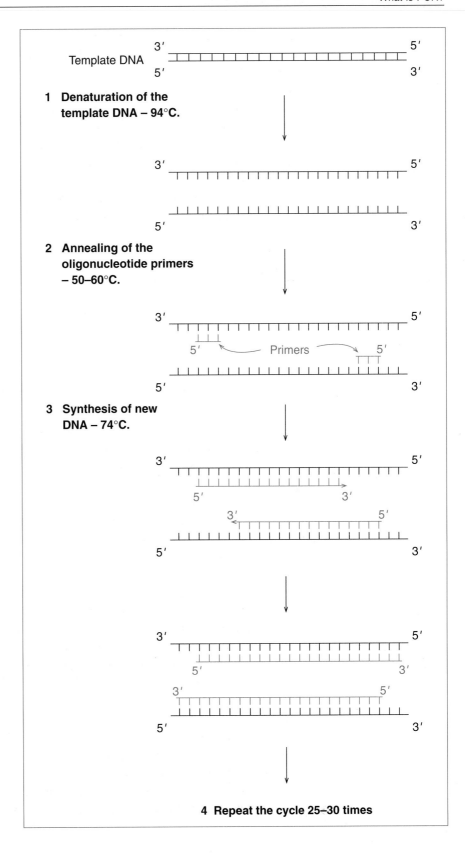

Template DNA

1 **Denaturation of the template DNA – 94°C.**

2 **Annealing of the oligonucleotide primers – 50–60°C.**

Primers

3 **Synthesis of new DNA – 74°C.**

4 **Repeat the cycle 25–30 times**

million new double-stranded molecules, each one a copy of the region of the starting molecule delineated by the annealing sites of the two primers.

1.5 Why gene cloning and PCR are so important

As you can see from Figures 1.1 and 1.2, gene cloning and PCR are relatively straightforward procedures. Why, then, have they assumed such importance in biology? The answer is largely because both techniques can provide a pure sample of an individual gene, separated from all the other genes in the cell.

1.5.1 Gene isolation by cloning

To understand exactly how cloning can provide a pure sample of a gene, consider the basic experiment from Figure 1.1, but drawn in a slightly different way (Figure 1.3). In this example the DNA fragment to be cloned is one member of a mixture of many different fragments, each carrying a different gene or part of a gene. This mixture could indeed be the entire genetic complement of an organism – a human, for instance. Each of these fragments becomes inserted into a different vector molecule to produce a family of recombinant DNA molecules, one of which carries the gene of interest. Usually only one recombinant DNA molecule is transported into any single host cell, so that although the final set of clones may contain many different recombinant DNA molecules, each individual clone contains multiple copies of just one molecule. The gene is now separated away from all the other genes in the original mixture, and its specific features can be studied in detail.

In practice, the key to the success or failure of a gene cloning experiment is the ability to identify the particular clone of interest from the many different ones that are obtained. If we consider the **genome** of the bacterium *Escherichia coli*, which contains just over 4000 different genes, we might at first despair of being able to find just one gene among all the possible clones (Figure 1.4). The problem becomes even more overwhelming when we remember that bacteria are relatively simple organisms and that the human genome contains about 10 times as many genes. However, as explained in Chapter 8, a variety of different strategies can be used to ensure that the correct gene can be obtained at the end of the cloning experiment. Some of these strategies involve modifications to the basic cloning procedure, so that only cells containing the desired recombinant DNA molecule can divide and the clone of interest is automatically **selected**. Other methods involve techniques that enable the desired clone to be identified from a mixture of lots of different clones.

Figure 1.3 Cloning allows individual fragments of DNA to be purified.

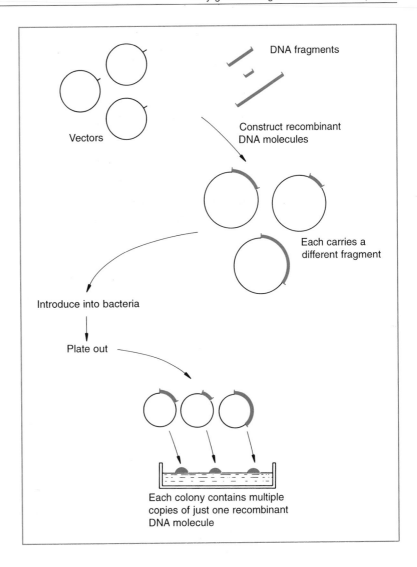

Once a gene has been cloned there is almost no limit to the information that can be obtained about the structure and expression of that gene. The availability of cloned material has stimulated the development of analytical methods for studying genes, with new techniques being introduced all the time. Methods for studying the structure and expression of a cloned gene are described in Chapters 10 and 11 respectively.

Gene isolation by PCR

The polymerase chain reaction can also be used to obtain a pure sample of a gene. This is because the region of the starting DNA molecule that is copied during PCR is the segment whose boundaries are marked by the annealing positions of the two oligonucleotide primers. If the primers anneal either side of the gene of interest, many copies of that gene will be synthesized

1.5.2

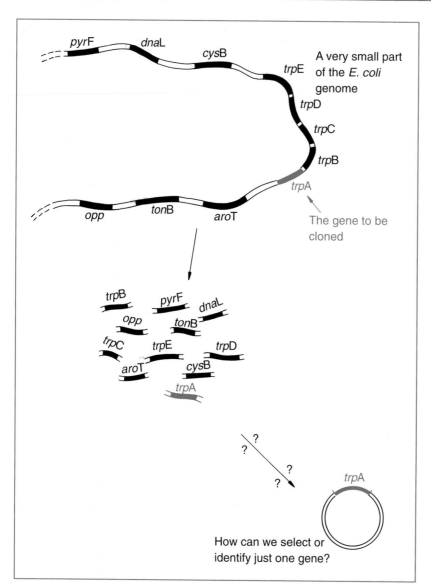

Figure 1.4 The problem of selection.

(Figure 1.5). The outcome is the same as with a gene cloning experiment, although the problem of selection does not arise because the desired gene is automatically 'selected' as a result of the positions at which the primers anneal.

A PCR experiment can be completed in a few hours, whereas it takes weeks if not months to obtain a gene by cloning. Why then is gene cloning still used? This is because of two limitations with PCR:

(1) In order for the primers to anneal to the correct positions, either side of the gene of interest, the sequences of these annealing sites must be known. It is easy to synthesize a primer with a predetermined sequence

Figure 1.5 Gene isolation by PCR.

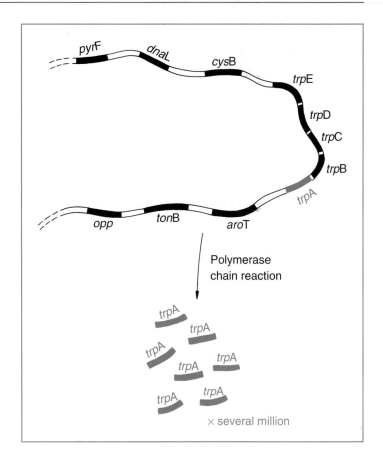

(see p. 171), but if the sequences of the annealing sites are unknown then the appropriate primers cannot be made. This means that PCR cannot be used to isolate genes that have not been studied before – that has to be done by cloning.

(2) There is a limit to the length of DNA sequence that can be copied by PCR. Five kilobases (kb) can be copied fairly easily, and segments up to 40 kb can be dealt with using specialized techniques, but this is shorter than the lengths of many genes, especially those of humans and other vertebrates. Cloning must be used if an intact version of a long gene is required.

Gene cloning is therefore the only way of isolating long genes or those that have never been studied before. But PCR still has many important applications. For example, even if the sequence of a gene is not known, it may still be possible to determine the appropriate sequences for a pair of primers, based on what is known about the sequence of the equivalent gene in a different organism. A gene that has been isolated and sequenced from, say, mouse could therefore be used to design a pair of primers for isolation of the equivalent gene from humans.

In addition, there are many applications where it is necessary to isolate or detect genes whose sequences are already known. A PCR of human globin genes, for example, is used to test for the presence of mutations that might cause the blood disease called thalassaemia. Design of appropriate primers for this PCR is easy because the sequences of the human globin genes are known. After the PCR, the gene copies are sequenced or studied in some other way to determine if any of the thalassaemia mutations are present.

Another clinical application of PCR involves the use of primers specific for the DNA of a disease-causing virus. A positive result indicates that a sample contains the virus and that the person who provided the sample should undergo treatment to prevent onset of the disease. The polymerase chain reaction is tremendously sensitive: a carefully set up reaction yields detectable amounts of DNA, even if there is just one DNA molecule in the starting mixture. This means that the technique can detect viruses at the earliest stages of an infection, increasing the chances of treatment being successful. This great sensitivity means that PCR can also be used with DNA from forensic material such as hairs and dried bloodstains or even from the bones of long-dead humans (Chapter 16).

1.6 How to find your way through this book

This book explains how gene cloning, PCR and other DNA analysis techniques are carried out and describes the applications of these techniques in modern biology. The applications are covered in the second and third parts of the book. Part 2 describes how genes and genomes are studied and Part 3 gives accounts of the broader applications of gene cloning and PCR in biotechnology, medicine, agriculture and forensic science.

In Part 1 we deal with the basic principles. Most of the nine chapters are devoted to gene cloning because this technique is more complicated than PCR. When you have understood how cloning is carried out you will have understood many of the basic principles of how DNA is analysed. In Chapter 2 we look at the central component of a gene cloning experiment – the vehicle – which transports the gene into the host cell and is responsible for its replication. To act as a cloning vehicle a DNA molecule must be capable of entering a host cell and, once inside, replicating to produce multiple copies of itself. Two naturally occurring types of DNA molecule satisfy these requirements:

(1) **Plasmids**, which are small circles of DNA found in bacteria and some other organisms. Plasmids can replicate independently of the host cell chromosome.

(2) **Virus chromosomes**, in particular the chromosomes of **bacteriophages**, which are viruses that specifically infect bacteria. During infection the bacteriophage DNA molecule is injected into the host cell where it undergoes replication.

Chapter 3 describes how DNA is purified from living cells – both the DNA that will be cloned and the vector DNA – and Chapter 4 covers the various techniques for handling purified DNA molecules in the laboratory. There are many such techniques, but two are particularly important in gene cloning. These are the ability to cut the vector at a specific point and then to repair it in such a way that the gene is inserted into the vehicle (Figure 1.1). These and other DNA manipulations were developed as an offshoot of basic research into DNA synthesis and modification within living cells, and most of the manipulations make use of purified enzymes. The properties of these enzymes, and the way they are used in DNA studies, are described in Chapter 4.

Once a recombinant DNA molecule has been constructed, it must be introduced into the host cell so that replication can take place. Transport into the host cell makes use of natural processes for uptake of plasmid and viral DNA molecules. These processes and the ways they are utilized in gene cloning are described in Chapter 5, and the most important types of cloning vector are introduced, and their uses examined, in Chapters 6 and 7. To conclude the coverage of gene cloning, in Chapter 8 we investigate the problem of selection (Figure 1.4), before returning in Chapter 9 to a more detailed description of PCR and its related techniques.

Further reading

Brock, T.D. (1990) *The Emergence of Bacterial Genetics*. Cold Spring Harbor Laboratory Press, New York. [Details the discovery of plasmids and bacteriophages.]

Brown, T.A. (1999) *Genomes*. BIOS Scientific Publishers, Oxford. [An introduction to genetics and molecular biology.]

Cherfas, J. (1982) *Man Made Life*. Blackwell, Oxford. [A history of the early years of genetic engineering.]

Judson, H.F. (1979) *The Eighth Day of Creation*. Penguin Science, London. [A very readable account of the development of molecular biology in the years before the gene cloning revolution.]

Mullis, K.B. (1990) The unusual origins of the polymerase chain reaction. *Scientific American*, **262** (4), 56–65. [An entertaining account of how PCR was invented.]

Turner, P.C., McLennan, A.G. & Bates, A.D. (2000) *Instant Notes in Molecular Biology*, 2nd edn. BIOS Scientific Publishers, Oxford. [A concise introduction to the principles of molecular biology.]

Chapter 2

Vehicles for Gene Cloning: Plasmids and Bacteriophages

Plasmids, 14   Bacteriophages, 19

A DNA molecule needs to display several features to be able to act as a vehicle for gene cloning. Most importantly it must be able to replicate within the host cell, so that numerous copies of the recombinant DNA molecule can be produced and passed to the daughter cells. A cloning vehicle also needs to be relatively small, ideally less than 10 kb in size, as large molecules tend to break down during purification, and are also more difficult to manipulate. Two kinds of DNA molecule that satisfy these criteria can be found in bacterial cells: plasmids and bacteriophage chromosomes. Although plasmids are frequently employed as cloning vehicles, two of the most important types of vector in use today are derived from bacteriophages.

2.1 Plasmids

2.1.1 Basic features of plasmids

Plasmids are circular molecules of DNA that lead an independent existence in the bacterial cell (Figure 2.1). Plasmids almost always carry one or more genes, and often these genes are responsible for a useful characteristic displayed by the host bacterium. For example, the ability to survive in normally toxic concentrations of antibiotics such as chloramphenicol or ampicillin is often due to the presence in the bacterium of a plasmid carrying antibiotic resistance genes. In the laboratory antibiotic resistance is often used as a **selectable marker** to ensure that bacteria in a culture contain a particular plasmid (Figure 2.2).

All plasmids possess at least one DNA sequence that can act as an **origin of replication**, so they are able to multiply within the cell quite independently of the main bacterial chromosome (Figure 2.3(a)). The smaller plasmids make use of the host cell's own DNA replicative enzymes in order to make copies of themselves, whereas some of the larger ones carry genes that code for special enzymes that are specific for plasmid replication.

Figure 2.1 Plasmids: independent genetic elements found in bacterial cells.

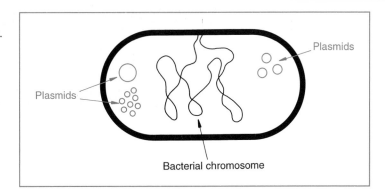

Figure 2.2 The use of antibiotic resistance as a selectable marker for a plasmid. RP4 (top) carries genes for resistance to ampicillin, tetracycline and kanamycin. Only those *E. coli* cells that contain RP4 (or a related plasmid) are able to survive and grow in a medium that contains toxic amounts of one or more of these antibiotics.

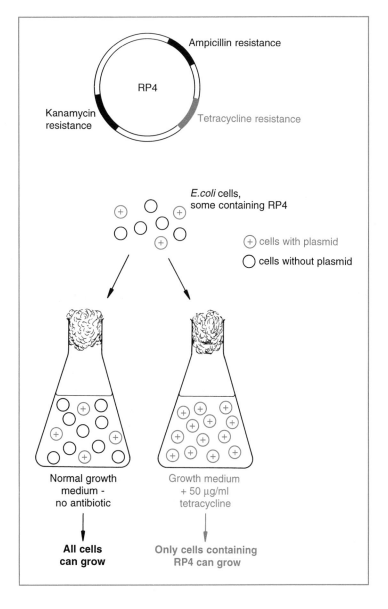

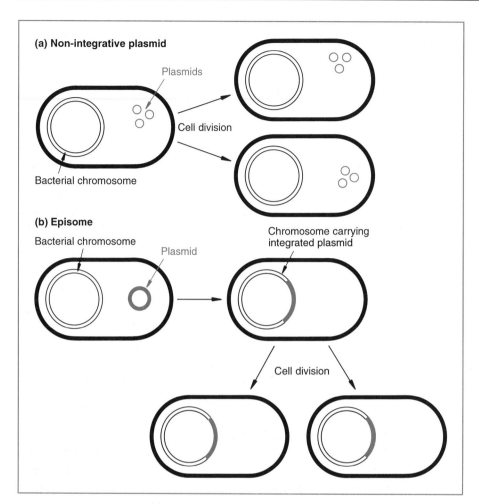

Figure 2.3
Replication strategies for (a) a non-integrative plasmid, and (b) an episome.

A few types of plasmid are also able to replicate by inserting themselves into the bacterial chromosome (Figure 2.3(b)). These integrative plasmids or **episomes** may be stably maintained in this form through numerous cell divisions, but will at some stage exist as independent elements. Integration is also an important feature of some bacteriophage chromosomes and will be described in more detail when these are considered (p. 20).

2.1.2 Size and copy number

The size and **copy number** of a plasmid are particularly important as far as cloning is concerned. We have already mentioned the relevance of plasmid size and stated that less than 10 kb is desirable for a cloning vehicle. Plasmids range from about 1.0 kb for the smallest to over 250 kb for the largest plasmids (Table 2.1), so only a few are useful for cloning purposes. However, as described in Chapter 7, larger plasmids may be adapted for cloning under some circumstances.

Table 2.1 Sizes of representative plasmids.

Plasmid	Size		Organism
	Nucleotide length (kb)	Molecular mass (MDa)	
pUC8	2.1	1.8	*E. coli*
ColE1	6.4	4.2	*E. coli*
RP4	54	36	*Pseudomonas* and others
F	95	63	*E. coli*
TOL	117	78	*Pseudomonas putida*
pTiAch5	213	142	*Agrobacterium tumefaciens*

The copy number refers to the number of molecules of an individual plasmid that are normally found in a single bacterial cell. The factors that control copy number are not well understood, but each plasmid has a characteristic value that may be as low as one (especially for the large molecules) or as many as 50 or more. Generally speaking, a useful cloning vehicle needs to be present in the cell in multiple copies so that large quantities of the recombinant DNA molecule can be obtained.

2.1.3 Conjugation and compatibility

Plasmids fall into two groups: conjugative and non-conjugative. Conjugative plasmids are characterized by the ability to promote sexual **conjugation** between bacterial cells (Figure 2.4), a process that can result in a conjugative plasmid spreading from one cell to all the other cells in a bacterial culture. Conjugation and plasmid transfer are controlled by a set of transfer or *tra* genes, which are present on conjugative plasmids but absent from the non-conjugative type. However, a non-conjugative plasmid may, under some circumstances, be cotransferred along with a conjugative plasmid when both are present in the same cell.

Several different kinds of plasmid may be found in a single cell, including more than one different conjugative plasmid at any one time. In fact, cells of *E. coli* have been known to contain up to seven different plasmids at once. To be able to coexist in the same cell, different plasmids must be **compatible**. If two plasmids are incompatible then one or the other will be quite rapidly lost from the cell. Different types of plasmid can therefore be assigned to different **incompatibility groups** on the basis of whether or not they can coexist, and plasmids from a single incompatibility group are often related to each other in various ways. The basis of incompatibility is not well understood, but events during plasmid replication are thought to underlie the phenomenon.

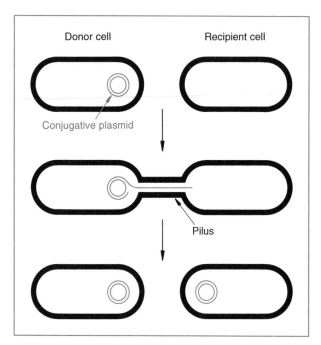

Donor cell

Recipient cell

Conjugative plasmid

Pilus

Figure 2.4 Plasmid transfer by conjugation between bacterial cells. The donor and recipient cells attach to each other by a pilus, a hollow appendage present on the surface of the donor cell. A copy of the plasmid is then passed to the recipient cell. Transfer is thought to occur through the pilus, but this has not been proven and transfer by some other means (e.g. directly across the bacterial cell walls) remains a possibility.

2.1.4 Plasmid classification

The most useful classification of naturally occurring plasmids is based on the main characteristic coded by the plasmid genes. The five main types of plasmid according to this classification are as follows:

(1) **Fertility** or **'F' plasmids** carry only *tra* genes and have no characteristic beyond the ability to promote conjugal transfer of plasmids, e.g. F plasmid of *E. coli*.

(2) **Resistance** or **'R' plasmids** carry genes conferring on the host bacterium resistance to one or more antibacterial agents, such as chloramphenicol, ampicillin and mercury. R plasmids are very important in clinical microbiology as their spread through natural populations can have profound consequences in the treatment of bacterial infections, e.g. RP4, commonly found in *Pseudomonas*, but also occurring in many other bacteria.

(3) **Col plasmids** code for colicins, proteins that kill other bacteria, e.g. ColE1 of *E. coli*.

(4) **Degradative plasmids** allow the host bacterium to metabolize unusual molecules such as toluene and salicylic acid, e.g. TOL of *Pseudomonas putida*.

(5) **Virulence plasmids** confer pathogenicity on the host bacterium; e.g. **Ti plasmids** of *Agrobacterium tumefaciens*, which induce crown gall disease on dicotyledonous plants.

2.1.5 Plasmids in organisms other than bacteria

Although plasmids are widespread in bacteria they are by no means as common in other organisms. The best characterized eukaryotic plasmid is the **2 μm circle** that occurs in many strains of the yeast *Saccharomyces cerevisiae*. The discovery of the 2 μm plasmid was very fortuitous as it has allowed the construction of vectors for cloning genes with this very important industrial organism as the host (p. 131). However, the search for plasmids in other eukaryotes (e.g. filamentous fungi, plants and animals) has proved disappointing, and it is suspected that many higher organisms simply do not harbour plasmids within their cells.

2.2 Bacteriophages

2.2.1 Basic features of bacteriophages

Bacteriophages, or phages as they are commonly known, are viruses that specifically infect bacteria. Like all viruses, phages are very simple in structure, consisting merely of a DNA (or occasionally ribonucleic acid (RNA)) molecule carrying a number of genes, including several for replication of the phage, surrounded by a protective coat or **capsid** made up of protein molecules (Figure 2.5).

The general pattern of infection, which is the same for all types of phage, is a three-step process (Figure 2.6).

(1)   The phage particle attaches to the outside of the bacterium and injects its DNA chromosome into the cell.
(2)   The phage DNA molecule is replicated, usually by specific phage enzymes coded by genes on the phage chromosome.

Figure 2.5 The two main types of phage structure: (a) head-and-tail (e.g. λ), and (b) filamentous (e.g. M13).

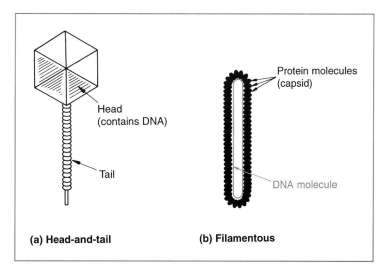

(a) Head-and-tail   **(b) Filamentous**

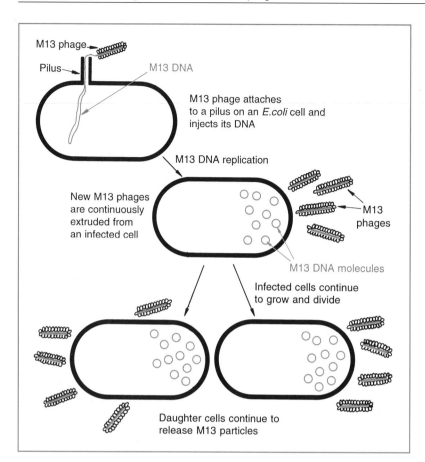

Figure 2.8 The infection cycle of bacteriophage M13.

M13 phage

Pilus

M13 DNA

M13 phage attaches to a pilus on an *E.coli* cell and injects its DNA

M13 DNA replication

New M13 phages are continuously extruded from an infected cell

M13 phages

M13 DNA molecules

Infected cells continue to grow and divide

Daughter cells continue to release M13 particles

phages, cell lysis never occurs, and the infected bacterium can continue to grow and divide, albeit at a slower rate than uninfected cells. Figure 2.8 shows the M13 infection cycle.

Although there are many different varieties of bacteriophage, only λ and M13 have found a major role as cloning vectors. The properties of these two phages will now be considered in more detail.

Gene organization in the λ DNA molecule

λ is a typical example of a head-and-tail phage (Figure 2.5(a)). The DNA is contained in the polyhedral head structure and the tail serves to attach the phage to the bacterial surface and to inject the DNA into the cell (Figure 2.7).

The λ DNA molecule is 49 kb in size and has been intensively studied by the techniques of gene mapping and **DNA sequencing**. As a result the positions and identities of most of the genes on the λ DNA molecule are known (Figure 2.9). A feature of the λ genetic map is that genes related in terms of function are clustered together on the genome. For example, all of the genes coding for components of the capsid are grouped together in the left-hand third of the molecule, and genes controlling integration of the prophage into

Figure 2.9 The λ genetic map, showing the positions of the important genes and the functions of the gene clusters.

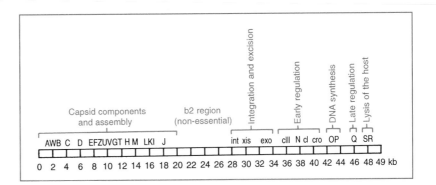

the host genome are clustered in the middle of the molecule. Clustering of related genes is profoundly important for controlling expression of the λ genome, as it allows genes to be switched on and off as a group rather than individually. Clustering is also important in the construction of λ-based cloning vectors (described in Chapter 6).

The linear and circular forms of λ DNA

A second feature of λ that turns out to be of importance in the construction of cloning vectors is the conformation of the DNA molecule. The molecule shown in Figure 2.9 is linear, with two free ends, and represents the DNA present in the phage head structure. This linear molecule consists of two **complementary** strands of DNA, base paired according to the **Watson–Crick rules** (that is, double-stranded DNA). However, at either end of the molecule is a short 12-nucleotide stretch, in which the DNA is single-stranded (Figure 2.10(a)). The two single strands are complementary, and so can base pair with one another to form a circular, completely double-stranded molecule (Figure 2.10(b)).

Complementary single strands are often referred to as **'sticky' ends** or cohesive ends, because base pairing between them can 'stick' together the two ends of a DNA molecule (or the ends of two different DNA molecules). The λ cohesive ends are called the ***cos* sites** and they play two distinct roles during the λ infection cycle. First, they allow the linear DNA molecule that is injected into the cell to be circularized, which is a necessary prerequisite for insertion into the bacterial genome (Figure 2.7).

The second role of the *cos* sites is rather different, and comes into play after the prophage has excised from the host genome. At this stage a large number of new λ DNA molecules are produced by the rolling circle mechanism of replication (Figure 2.10(c)), in which a continuous DNA strand is 'rolled off' the template molecule. The result is a catenane consisting of a series of linear λ genomes joined together at the *cos* sites. The role of the *cos* sites is now to act as recognition sequences for an **endonuclease** that cleaves

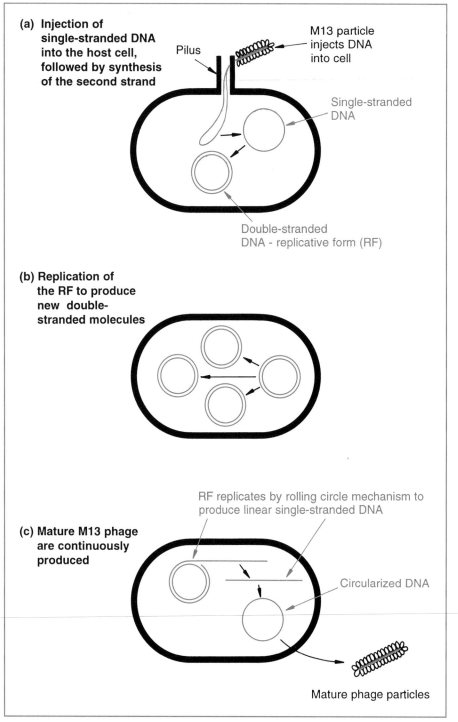

(a) Injection of single-stranded DNA into the host cell, followed by synthesis of the second strand

Pilus

M13 particle injects DNA into cell

Single-stranded DNA

Double-stranded DNA - replicative form (RF)

(b) Replication of the RF to produce new double-stranded molecules

RF replicates by rolling circle mechanism to produce linear single-stranded DNA

(c) Mature M13 phage are continuously produced

Circularized DNA

Mature phage particles

Figure 2.11 The M13 infection cycle, showing the different types of DNA replication that occur. (a) After infection the single-stranded M13 DNA molecule is converted into the double-stranded replicative form (RF). (b) The RF replicates to produce multiple copies of itself. (c) Single-stranded molecules are synthesized by rolling circle replication and used in the assembly of new M13 particles.

remembered that plasmids are not commonly found in organisms other than bacteria and yeast (p. 19).

In fact viruses have considerable potential as cloning vehicles for some types of applications with animal cells. Mammalian viruses such as **simian virus 40 (SV40)**, **adenoviruses** and **retroviruses**, and the insect **baculoviruses**, are the ones that have received most attention so far. These are discussed more fully in Chapter 7.

Further reading

Dale, J.W. (1998) *Molecular Genetics of Bacteria*, 3rd edn. Chichester, John Wiley. [Provides a detailed description of plasmids and bacteriophages.]

Prescott, L.M., Harley, J.P. & Klein, D.A. (2000) *Microbiology*, 4th edn. William C. Brown, Dubuque. [A good introduction to microbiology, including plasmids and phages.]

Purification of DNA from Living Cells

The genetic engineer will, at different times, need to prepare at least three distinct kinds of DNA. Firstly, **total cell DNA** will often be required as a source of material from which to obtain genes to be cloned. Total cell DNA may be DNA from a culture of bacteria, from a plant, from animal cells, or from any other type of organism that is being studied.

The second type of DNA that will be required is pure plasmid DNA. Preparation of plasmid DNA from a culture of bacteria follows the same basic steps as purification of total cell DNA, with the crucial difference that at some stage the plasmid DNA must be separated from the main bulk of chromosomal DNA also present in the cell.

Finally, phage DNA will be needed if a phage cloning vehicle is to be used. Phage DNA is generally prepared from bacteriophage particles rather than from infected cells, so there is no problem with contaminating bacterial DNA. However, special techniques are needed to remove the phage capsid. An exception is the double-stranded replicative form of M13, which is prepared from *E. coli* cells in the same way as a bacterial plasmid.

3.1 Preparation of total cell DNA

The fundamentals of DNA preparation are most easily understood by first considering the simplest type of DNA purification procedure, that where the entire DNA complement of a bacterial cell is required. The modifications needed for plasmid and phage DNA preparation can then be described later.

The procedure for total DNA preparation from a culture of bacterial cells can be divided into four stages (Figure 3.1).

(1) A culture of bacteria is grown and then **harvested.**
(2) The cells are broken open to release their contents.
(3) This **cell extract** is treated to remove all components except the DNA.
(4) The resulting DNA solution is concentrated.

Figure 3.1 The basic steps in preparation of total cell DNA from a culture of bacteria.

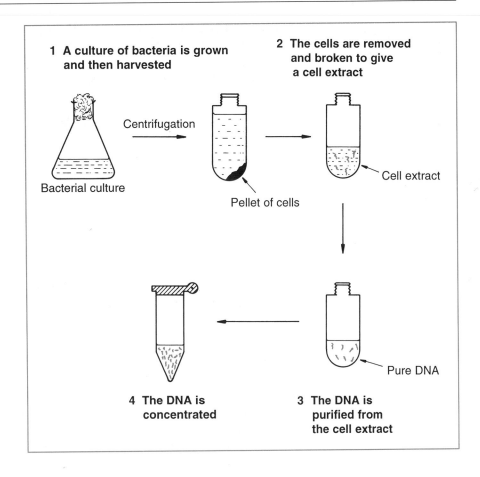

1 **A culture of bacteria is grown and then harvested**

2 **The cells are removed and broken to give a cell extract**

Centrifugation

Bacterial culture

Pellet of cells

Cell extract

Pure DNA

4 **The DNA is concentrated**

3 **The DNA is purified from the cell extract**

Growing and harvesting a bacterial culture

Most bacteria can be grown without too much difficulty in a liquid medium (**broth culture**). The culture medium must provide a balanced mixture of the essential nutrients at concentrations that will allow the bacteria to grow and divide efficiently. Two typical growth media are detailed in Table 3.1.

M9 is an example of a **defined medium** in which all the components are known. This medium contains a mixture of inorganic nutrients to provide essential elements such as nitrogen, magnesium and calcium, as well as glucose to supply carbon and energy. In practice, additional growth factors such as trace elements and vitamins must be added to M9 before it will support bacterial growth. Precisely which supplements are needed depends on the species concerned.

The second medium described in Table 3.1 is rather different. Luria-Bertani (LB) is a complex or **undefined medium**, meaning that the precise identity and quantity of its components are not known. This is because two of the ingredients, tryptone and yeast extract, are complicated mixtures of unknown chemical compounds. Tryptone in fact supplies amino acids and small peptides, while yeast extract (a dried preparation of partially digested

Table 3.1 The composition of two typical media for the growth of bacterial cultures.

Medium	Component	g/l of medium
M9 medium	Na_2HPO_4	6.0
	KH_2PO_4	3.0
	NaCl	0.5
	NH_4Cl	1.0
	$MgSO_4$	0.5
	Glucose	2.0
	$CaCl_2$	0.015
LB (Luria-Bertani medium)	Tryptone	10
	Yeast extract	5
	NaCl	10

yeast cells) provides the nitrogen requirements, along with sugars and inorganic and organic nutrients. Complex media such as LB need no further supplementation and support the growth of a wide range of bacterial species.

Defined media must be used when the bacterial culture has to be grown under precisely controlled conditions. However, this is not necessary when the culture is being grown simply as a source of DNA, and under these circumstances a complex medium is appropriate. In LB medium at 37°C, aerated by shaking at 150–250 rpm on a rotary platform, *E. coli* cells divide once every 20 min or so until the culture reaches a maximum density of about $2–3 \times 10^9$ cells/ml. The growth of the culture can be monitored by reading the optical density (OD) at 600 nm (Figure 3.2), at which wavelength one OD unit corresponds to about 0.8×10^9 cells/ml.

In order to prepare a cell extract, the bacteria must be obtained in as small a volume as possible. Harvesting is therefore performed by spinning the culture in a centrifuge (Figure 3.3). Fairly low centrifugation speeds will pellet the bacteria at the bottom of the centrifuge tube, allowing the culture medium to be poured off. Bacteria from a 1000 ml culture at maximum cell density can then be resuspended into a volume of 10 ml or less.

3.1.2   Preparation of a cell extract

The bacterial cell is enclosed in a cytoplasmic membrane and surrounded by a rigid cell wall. With some species, including *E. coli*, the cell wall may itself be enveloped by a second, outer membrane. All of these barriers have to be disrupted to release the cell components.

Techniques for breaking open bacterial cells can be divided into physical methods, in which the cells are disrupted by mechanical forces, and chemical methods, where cell lysis is brought about by exposure to chemical agents that affect the integrity of the cell barriers. Chemical methods are most commonly used with bacterial cells when the object is DNA preparation.

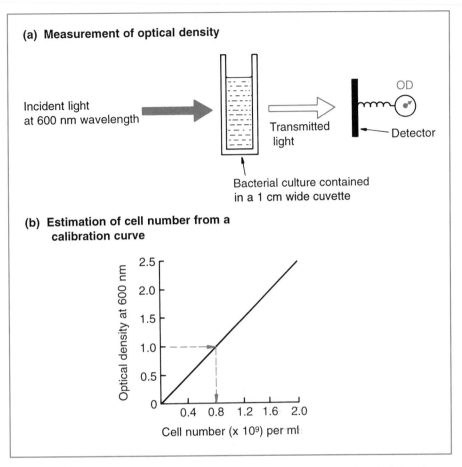

(a) Measurement of optical density

Incident light
at 600 nm wavelength

Transmitted
light

OD

Detector

Bacterial culture contained
in a 1 cm wide cuvette

(b) Estimation of cell number from a
   calibration curve

Optical density at 600 nm

2.5
2.0
1.5
1.0
0.5
0

0.4 0.8 1.2 1.6 2.0

Cell number (x 10^9) per ml

Figure 3.2 Estimation of bacterial cell number by measurement of optical density (OD). (a) A sample of the culture is placed in a glass cuvette and light with a wavelength of 600 nm shone through. The amount of light that passes through the culture is measured and the OD (also called the absorbance) calculated as 1 OD unit = $-\log_{10}$ (intensity of transmitted light)/(intensity of incident light). The operation is performed with a spectrophotometer. (b) The cell number corresponding to the OD reading is calculated from a calibration curve. This curve is plotted from the OD values of a series of cultures of known cell density. For *E. coli*, 1 OD unit = 0.8×10^9 cells/ml.

Chemical lysis generally involves one agent attacking the cell wall and another disrupting the cell membrane (Figure 3.4(a)). The chemicals that are used depend on the species of bacterium involved, but with *E. coli* and related organisms, weakening of the cell wall is usually brought about by **lysozyme**, ethylenediamine tetraacetate (EDTA), or a combination of both. Lysozyme is an enzyme that is present in egg white and in secretions such as tears and saliva, and which digests the polymeric compounds that give the cell wall its rigidity. On the other hand, EDTA removes magnesium ions that

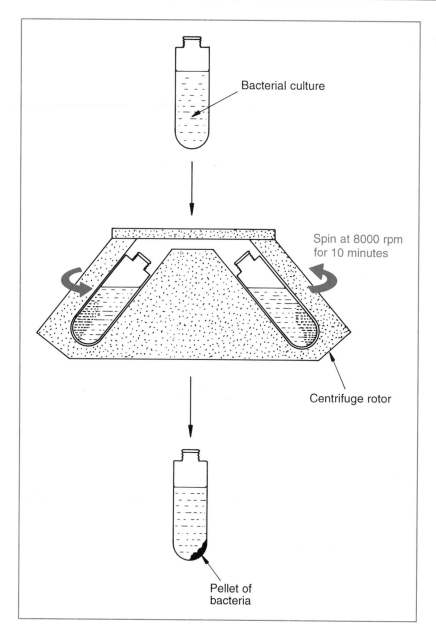

Figure 3.3 Harvesting bacteria by centrifugation.

Bacterial culture

Spin at 8000 rpm for 10 minutes

Centrifuge rotor

Pellet of bacteria

are essential for preserving the overall structure of the cell envelope, and also inhibits cellular enzymes that could degrade DNA. Under some conditions, weakening the cell wall with lysozyme or EDTA is sufficient to cause bacterial cells to burst, but usually a detergent such as sodium dodecyl sulphate (SDS) is also added. Detergents aid the process of lysis by removing lipid molecules and thereby cause disruption of the cell membranes.

Having lysed the cells, the final step in preparation of a cell extract is removal of insoluble cell debris. Components such as partially digested cell

Figure 3.4 Preparation of a cell extract: (a) cell lysis, and (b) centrifugation of the cell extract to remove insoluble debris.

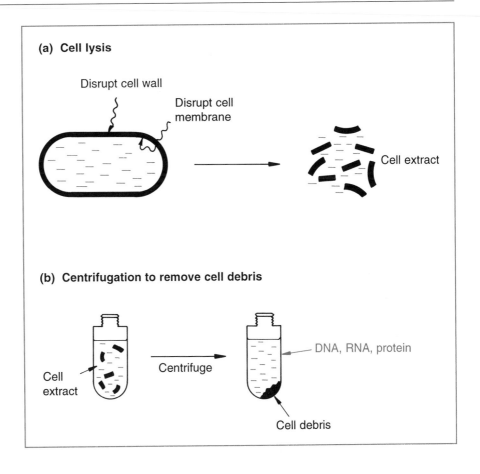

(a) **Cell lysis**

Disrupt cell wall

Disrupt cell membrane

Cell extract

(b) **Centrifugation to remove cell debris**

Cell extract

Centrifuge

DNA, RNA, protein

Cell debris

wall fractions can be pelleted by centrifugation (Figure 3.4(b)), leaving the cell extract as a reasonably clear supernatant.

Purification of DNA from a cell extract

In addition to DNA, a bacterial cell extract contains significant quantities of protein and RNA. A variety of procedures can be used to remove these contaminants, leaving the DNA in a pure form.

The standard way to deproteinize a cell extract is to add phenol or a 1:1 mixture of phenol and chloroform. These organic solvents precipitate proteins but leave the nucleic acids (DNA and RNA) in aqueous solution. The result is that if the cell extract is mixed gently with the solvent, and the layers then separated by centrifugation, precipitated protein molecules are left as a white coagulated mass at the interface between the aqueous and organic layers (Figure 3.5). The aqueous solution of nucleic acids can then be removed with a pipette.

With some cell extracts the protein content is so great that a single phenol extraction is not sufficient to purify completely the nucleic acids. This problem could be solved by carrying out several phenol extractions one after

3.1.3

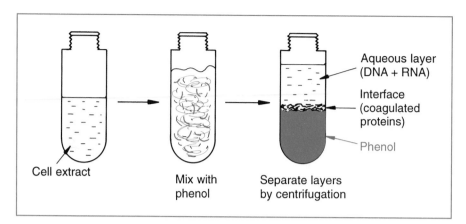

Figure 3.5 Removal of protein contaminants by phenol extraction.

Aqueous layer
(DNA + RNA)

Interface
(coagulated
proteins)

Phenol

Cell extract

Mix with
phenol

Separate layers
by centrifugation

the other, but this is undesirable as each mixing and centrifugation step results in a certain amount of breakage of the DNA molecules. The answer is to treat the cell extract with a **protease** such as pronase or proteinase K before phenol extraction. These enzymes break polypeptides down into smaller units, which are more easily removed by phenol.

Some RNA molecules, especially **messenger RNA (mRNA)**, are removed by phenol treatment, but most remain with the DNA in the aqueous layer. The only effective way to remove the RNA is with the enzyme **ribonuclease**, which rapidly degrades these molecules into ribonucleotide subunits.

3.1.4   Concentration of DNA samples

Often a successful preparation results in a very thick solution of DNA that does not need to be concentrated any further. However, dilute solutions are sometimes obtained and it is important to consider methods for increasing the DNA concentration.

The most frequently used method of concentration is **ethanol precipitation**. In the presence of salt (strictly speaking, monovalent cations such as sodium ions (Na^+)), and at a temperature of $-20°C$ or less, absolute ethanol efficiently precipitates polymeric nucleic acids. With a thick solution of DNA the ethanol can be layered on top of the sample, causing molecules to precipitate at the interface. A spectacular trick is to push a glass rod through the ethanol into the DNA solution. When the rod is removed, DNA molecules adhere and can be pulled out of the solution in the form of a long fibre (Figure 3.6(a)). Alternatively, if ethanol is mixed with a dilute DNA solution, the precipitate can be collected by centrifugation (Figure 3.6(b)), and then redissolved in an appropriate volume of water. Ethanol precipitation has the added advantage of leaving short-chain and monomeric nucleic acid components in solution. Ribonucleotides produced by ribonuclease treatment are therefore lost at this stage.

Figure 3.6 Collecting DNA by ethanol precipitation. (a) Absolute ethanol is layered on top of a concentrated solution of DNA. Fibres of DNA can be withdrawn with a glass rod. (b) For less concentrated solutions ethanol is added (at a ratio of 2.5 volumes of absolute ethanol to 1 volume of DNA solution) and precipitated DNA collected by centrifugation.

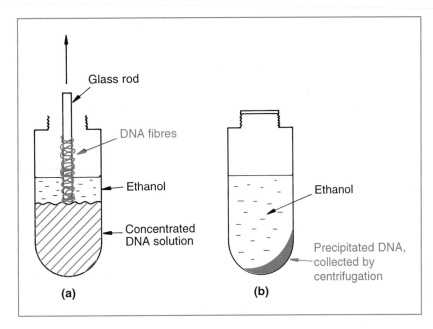

Glass rod

DNA fibres

Ethanol

Concentrated
DNA solution

Ethanol

Precipitated DNA,
collected by
centrifugation

(a)

(b)

3.1.5 Measurement of DNA concentration

It is crucial to know exactly how much DNA is present in a solution when carrying out a gene cloning experiment. Fortunately DNA concentrations can be accurately measured by **ultraviolet (UV) absorbance spectrophotometry**. The amount of UV radiation absorbed by a solution of DNA is directly proportional to the amount of DNA in the sample. Usually absorbance is measured at 260 nm, at which wavelength an absorbance (A_{260}) of 1.0 corresponds to 50 µg of double-stranded DNA per ml.

Ultraviolet absorbance can also be used to check the purity of a DNA preparation. With a pure sample of DNA the ratio of the absorbances at 260 nm and 280 nm (A_{260}/A_{280}) is 1.8. Ratios of less than 1.8 indicate that the preparation is contaminated, either with protein or with phenol.

3.1.6 Preparation of total cell DNA from organisms other than bacteria

Bacteria are not the only organisms from which DNA may be required. Total cell DNA from, for example, from plants or animals will be needed if the aim of the genetic engineering project is to clone genes from these organisms. Although the basic steps in DNA purification are the same whatever the organism, some modifications may have to be introduced to take account of the special features of the cells being used.

Obviously growth of cells in liquid medium may not always be appropriate, even though plant and animal cell cultures are becoming increasingly important in biology. The major modifications, however, are likely to be needed at the cell breakage stage. The chemicals used for disrupting bacterial cells do not usually work with other organisms: lysozyme, for example, has no effect on plant cells. Specific degradative enzymes are available for most cell wall types, but often physical techniques, such as grinding frozen material with a mortar and pestle, are more efficient. On the other hand, most animal cells have no cell wall at all, and can be lysed simply by treating with detergent.

Another important consideration is the biochemical content of the cells from which DNA is being extracted. With most bacteria the main biochemicals present in a cell extract are protein, DNA and RNA, so phenol extraction and/or protease treatment, followed by removal of RNA with ribonuclease, leaves a pure DNA sample. These treatments may not, however, be sufficient to give pure DNA if the cells also contain significant quantities of other biochemicals. Plant tissues are particularly difficult in this respect as they often contain large amounts of carbohydrates that are not removed by phenol extraction. Instead a different approach must be used.

One method makes use of a detergent called cetyltrimethylammonium bromide (CTAB), which forms an insoluble complex with nucleic acids. When CTAB is added to a plant cell extract the nucleic acid–CTAB complex precipitates, leaving carbohydrate, protein and other contaminants in the supernatant (Figure 3.7). The precipitate is then collected by centrifugation and resuspended in 1 M NaCl (sodium chloride), which causes the complex to break down. The nucleic acids can now be concentrated by ethanol precipitation and the RNA removed by ribonuclease treatment.

A second method involves a compound called guanidinium thiocyanate, which has two properties that make it useful for DNA purification. First, it denatures and dissolves all biochemicals other than nucleic acids and can therefore be used to release DNA from virtually any type of tissue. Secondly, in the presence of guanidinium thiocyanate, DNA binds tightly to silica particles (Figure 3.8(a)). This provides an easy way of recovering the DNA from the denatured mix of biochemicals. One possibility is to add the silica directly to the cell extract, but it is more convenient to use a chromatography column. The silica is placed in the column and the cell extract added (Figure 3.8(b)). DNA binds to the silica and is retained in the column, whereas the denatured biochemicals pass straight through. After washing away the last contaminants with guanidinium thiocyanate solution, the DNA is recovered by adding water, which destabilizes the interactions between the DNA molecules and the silica.

3.2   Preparation of plasmid DNA

Purification of plasmids from a culture of bacteria involves the same general strategy as preparation of total cell DNA. A culture of cells, containing plasmids, is grown in liquid medium, harvested, and a cell extract prepared.

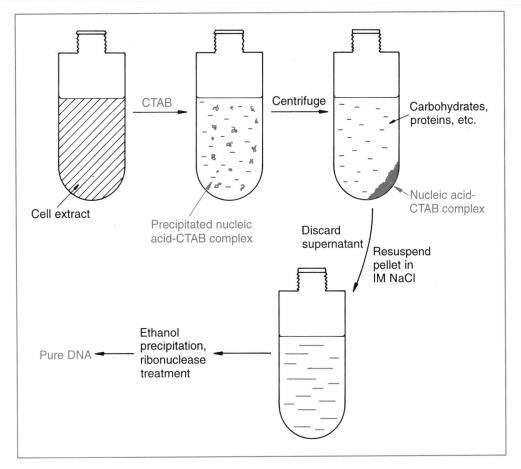

Figure 3.7 The CTAB method for purification of plant DNA.

The extract is deproteinized, the RNA removed, and the DNA probably concentrated by ethanol precipitation. However, there is an important distinction between plasmid purification and preparation of total cell DNA. In a plasmid preparation it is always necessary to separate the plasmid DNA from the large amount of bacterial chromosomal DNA that is also present in the cells.

Separating the two types of DNA can be very difficult, but is nonetheless essential if the plasmids are to be used as cloning vehicles. The presence of the smallest amount of contaminating bacterial DNA in a gene cloning experiment may easily lead to undesirable results. Fortunately several methods are available for removal of bacterial DNA during plasmid purification, and the use of these methods, individually or in combination, can result in isolation of very pure plasmid DNA.

The methods are based on the several physical differences between plasmid DNA and bacterial DNA, the most obvious of which is size. The largest plasmids are only 8% of the size of the *E. coli* chromosome, and most

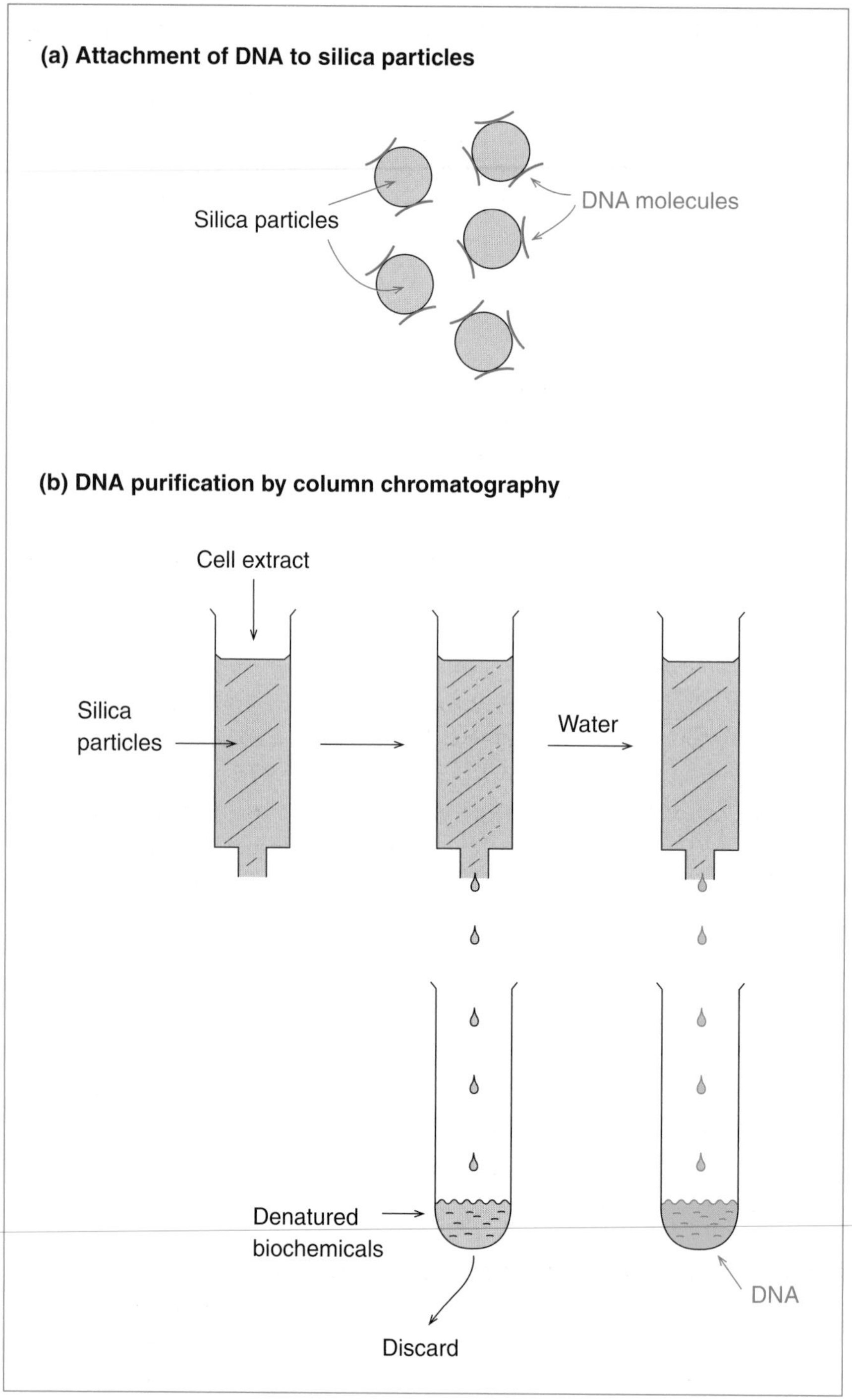

(a) Attachment of DNA to silica particles

Silica particles

DNA molecules

(b) DNA purification by column chromatography

Cell extract

Silica particles

Water

Denatured biochemicals

Discard

DNA

Figure 3.8 DNA purification by the guanidinium thiocyanate and silica method. (a) In the presence of guanidinium thiocyanate, DNA binds to silica particles. (b) DNA is purified by column chromatography.

are much smaller than this. Techniques that can separate small DNA molecules from large ones should therefore effectively purify plasmid DNA.

In addition to size, plasmids and bacterial DNA differ in **conformation**. When applied to a polymer such as DNA, the term conformation refers to the overall spatial configuration of the molecule, with the two simplest conformations being linear and circular. Plasmids and the bacterial chromosome are circular, but during preparation of the cell extract the chromosome is always broken to give linear fragments. A method for separating circular from linear molecules will therefore result in pure plasmids.

3.2.1   Separation on the basis of size

The usual stage at which size fractionation is performed is during preparation of the cell extract. If the cells are lysed under very carefully controlled conditions, only a minimal amount of chromosomal DNA breakage occurs. The resulting DNA fragments are still very large – much larger than the plasmids – and can be removed with the cell debris by centrifugation. This process is aided by the fact that the bacterial chromosome is physically attached to the cell envelope, so fragments of the chromosome sediment with the cell debris if these attachments are not broken.

Cell disruption must therefore be carried out very gently to prevent wholesale breakage of the bacterial DNA. For *E. coli* and related species, controlled lysis is performed as shown in Figure 3.9. Treatment with EDTA and lysozyme is carried out in the presence of sucrose, which prevents the cells from bursting immediately. Instead, **sphaeroplasts** are formed, cells with partially degraded cell walls that retain an intact cytoplasmic membrane. Cell lysis is now induced by adding a non-ionic detergent such as Triton X-100 (ionic detergents, such as SDS, cause chromosomal breakage). This method causes very little breakage of the bacterial DNA, so centrifugation leaves a **cleared lysate**, consisting almost entirely of plasmid DNA.

A cleared lysate will, however, invariably retain some chromosomal DNA. Furthermore, if the plasmids themselves are large molecules, they may also sediment with the cell debris. Size fractionation is therefore rarely sufficient on its own, and we must consider alternative ways of removing the bacterial DNA contaminants.

3.2.2   Separation on the basis of conformation

Before considering the ways in which conformational differences between plasmids and bacterial DNA can be used to separate the two types of DNA, we must look more closely at the overall structure of plasmid DNA. It is not strictly correct to say that plasmids have a circular conformation, because double-stranded DNA circles can in fact take up one of two quite distinct configurations. Most plasmids exist in the cell as **supercoiled** molecules (Figure 3.10(a)). Supercoiling occurs because the double helix of the plasmid DNA is partially unwound during the plasmid replication process by

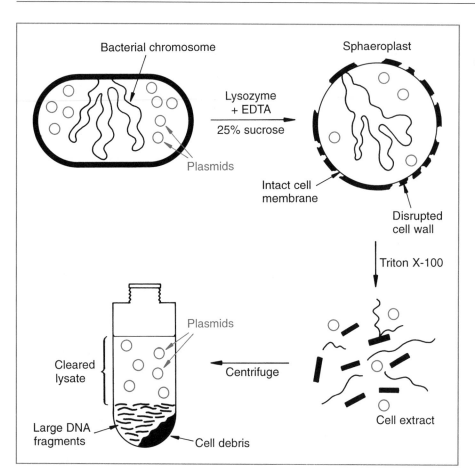

Figure 3.9
Preparation of a cleared lysate.

enzymes called topoisomerases (p. 59). The supercoiled conformation can be maintained only if both polynucleotide strands are intact, hence the more technical name of **covalently closed-circular (ccc) DNA**. If one of the poly-nucleotide strands is broken the double helix reverts to its normal **relaxed** state, and the plasmid takes on the alternative conformation, called **open-circular (oc)** (Figure 3.10(b)).

Supercoiling is important in plasmid preparation because supercoiled molecules can be fairly easily separated from non-supercoiled DNA. Two dif-ferent methods are commonly used. Both can purify plasmid DNA from crude cell extracts, although in practice best results are obtained if a cleared lysate is first prepared.

Alkaline denaturation

The basis of this technique is that there is a narrow pH range at which non-supercoiled DNA is denatured, whereas supercoiled plasmids are not. If sodium hydroxide is added to a cell extract or cleared lysate, so that the pH

Figure 3.10 Two
conformations of circular
double-stranded DNA: (a)
supercoiled – both strands
are intact; (b) open-circular –
one or both strands are
nicked.

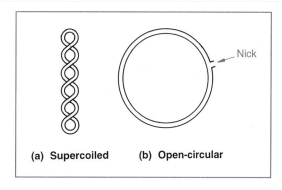

(a) Supercoiled   **(b) Open-circular**

Figure 3.11 Plasmid
purification by the alkaline
denaturation method.

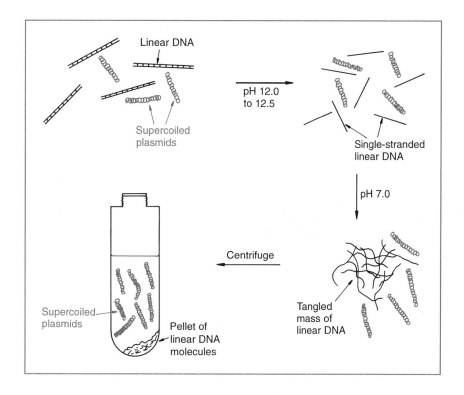

is adjusted to 12.0–12.5, then the hydrogen bonding in non-supercoiled DNA
molecules is broken, causing the double helix to unwind and the two polynucleotide chains to separate (Figure 3.11). If acid is now added, these denatured bacterial DNA strands reaggregate into a tangled mass. The insoluble
network can be pelleted by centrifugation, leaving pure plasmid DNA in the
supernatant.

   An additional advantage of this procedure is that, under some circumstances (specifically cell lysis by SDS and neutralization with sodium
acetate), most of the protein and RNA also becomes insoluble and can be
removed by the centrifugation step. Phenol extraction and ribonuclease

treatment may therefore not be needed if the alkaline denaturation method is used.

Ethidium bromide–caesium chloride density gradient centrifugation

This is a specialized version of the more general technique of equilibrium or **density gradient centrifugation**. A density gradient is produced by centrifuging a solution of caesium chloride (CsCl) at a very high speed (Figure 3.12(a)). The gradient develops because the high centrifugal force pulls the caesium and chloride ions towards the bottom of the tube. Their downward migration is counterbalanced by diffusion, so a concentration gradient is set up, with the CsCl density greater towards the bottom of the tube.

Macromolecules present in the CsCl solution when it is centrifuged form bands at distinct points in the gradient (Figure 3.12(b)). Exactly where a particular molecule bands depends on its **buoyant density**; DNA has a buoyant density of about $1.7\,g/cm^3$, and therefore migrates to the point in the gradient where the CsCl density is also $1.7\,g/cm^3$. In contrast, protein molecules have much lower buoyant densities, and so float at the top of the tube, whereas RNA forms a pellet at the bottom (Figure 3.12(b)). Density gradient centrifugation can therefore separate DNA, RNA and protein and is an alternative to phenol extraction and ribonuclease treatment for DNA purification.

More importantly, density gradient centrifugation in the presence of **ethidium bromide (EtBr)** can be used to separate supercoiled DNA from non-supercoiled molecules. Ethidium bromide binds to DNA molecules by

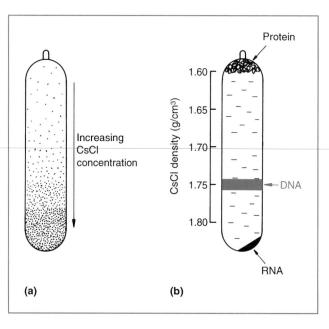

Figure 3.12 Caesium chloride density gradient centrifugation. (a) A CsCl density gradient produced by high speed centrifugation. (b) Separation of protein, DNA and RNA in a density gradient.

Figure 3.13 Partial unwinding of the DNA double helix by EtBr intercalation between adjacent base pairs. The normal DNA molecule shown on the left is partially unwound by taking up four EtBr molecules, resulting in the 'stretched' structure on the right.

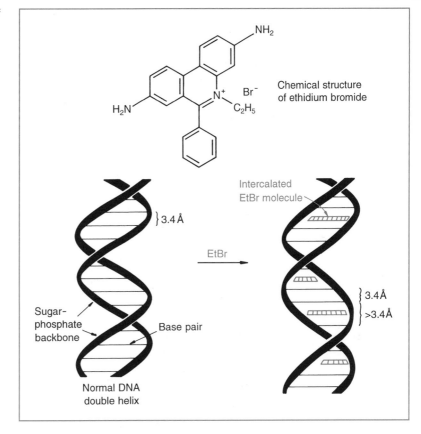

Chemical structure of ethidium bromide

Intercalated EtBr molecule

3.4 Å

EtBr

3.4 Å
>3.4 Å

Sugar–phosphate backbone

Base pair

Normal DNA double helix

intercalating between adjacent base pairs, causing partial unwinding of the double helix (Figure 3.13). This unwinding results in a decrease in the buoyant density, by as much as $0.125\,g/cm^3$ for linear DNA. However supercoiled DNA, with no free ends, has very little freedom to unwind, and can only bind a limited amount of EtBr. The decrease in buoyant density of a supercoiled molecule is therefore much less, only about $0.085\,g/cm^3$. As a consequence, supercoiled molecules form a band in an EtBr–CsCl gradient at a different position to linear and open-circular DNA (Figure 3.14(a)).

Ethidium bromide–caesium chloride density gradient centrifugation is a very efficient method for obtaining pure plasmid DNA. When a cleared lysate is subjected to this procedure, plasmids band at a distinct point, separated from the linear bacterial DNA, with the protein floating on the top of the gradient and RNA pelleted at the bottom. The position of the DNA bands can be seen by shining ultraviolet radiation on the tube, which causes the bound EtBr to fluoresce. The pure plasmid DNA is removed by puncturing the side of the tube and withdrawing a sample with a syringe (Figure 3.14(b)). The EtBr bound to the plasmid DNA is extracted with *n*-butanol (Figure 3.14(c)) and the CsCl removed by dialysis (Figure 3.14(d)). The resulting plasmid preparation is virtually 100% pure and ready for use as a cloning vehicle.

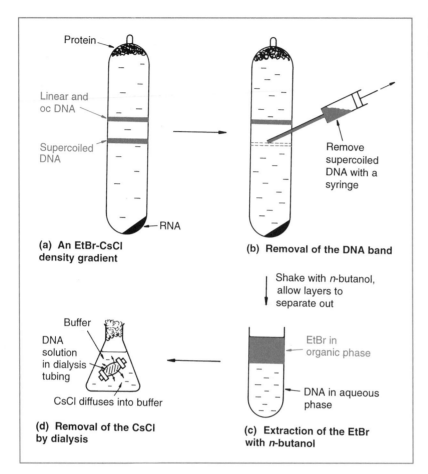

Figure 3.14 Purification of plasmid DNA by EtBr–CsCl density gradient centrifugation.

(a) An EtBr-CsCl density gradient

Protein
Linear and oc DNA
Supercoiled DNA
RNA

(b) Removal of the DNA band

Remove supercoiled DNA with a syringe

Shake with *n*-butanol, allow layers to separate out

(c) Extraction of the EtBr with *n*-butanol

EtBr in organic phase
DNA in aqueous phase

(d) Removal of the CsCl by dialysis

Buffer
DNA solution in dialysis tubing
CsCl diffuses into buffer

3.2.3 Plasmid amplification

Preparation of plasmid DNA can be hindered by the fact that plasmids make up only a small proportion of the total DNA in the bacterial cell. The yield of DNA from a bacterial culture may therefore be disappointingly low. **Plasmid amplification** offers a means of increasing this yield.

The aim of amplification is to increase the copy number of a plasmid. Some **multicopy plasmids** (those with copy numbers of 20 or more) have the useful property of being able to replicate in the absence of protein synthesis. This contrasts with the main bacterial chromosome, which cannot replicate under these conditions. This property can be made use of during the growth of a bacterial culture for plasmid DNA purification. After a satisfactory cell density has been reached, an inhibitor of protein synthesis (e.g. chloramphenicol) is added, and the culture incubated for a further 12 hours. During this time the plasmid molecules continue to replicate, even though chromosome replication and cell division are blocked (Figure 3.15). The result is that plasmid copy numbers of several thousand may be attained. Amplification is therefore a very efficient way of increasing the yield of multicopy plasmids.

Figure 3.15 Plasmid amplification.

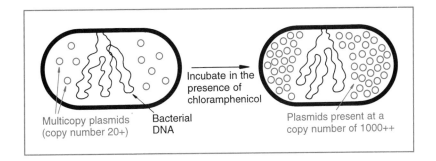

Multicopy plasmids (copy number 20+)

Bacterial DNA

Incubate in the presence of chloramphenicol

Plasmids present at a copy number of 1000++

Figure 3.16 Preparation of a phage suspension from an infected culture of bacteria.

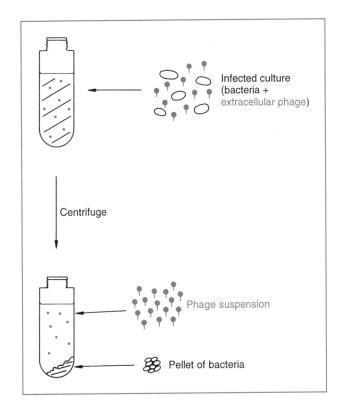

Infected culture (bacteria + extracellular phage)

Centrifuge

Phage suspension

Pellet of bacteria

3.3

Preparation of bacteriophage DNA

The key difference between phage DNA purification and the preparation of either total cell DNA or plasmid DNA is that for phages the starting material is not normally a cell extract. This is because bacteriophage particles can be obtained in large numbers from the extracellular medium of an infected bacterial culture. When such a culture is centrifuged, the bacteria are pelleted, leaving the phage particles in suspension (Figure 3.16). The phage particles are then collected from the suspension and their DNA extracted by a single deproteinization step to remove the phage capsid.

This overall process is rather more straightforward than the procedure used to prepare total cell or plasmid DNA. Nevertheless, successful purification of significant quantities of phage DNA is subject to several pitfalls. The main difficulty, especially with λ, is growing an infected culture in such a way that the extracellular phage titre (the number of phage particles per ml of culture) is sufficiently high. In practical terms, the maximum titre that can reasonably be expected for λ is 10^{10} per ml; yet 10^{10} λ particles will yield only 500 ng of DNA. Large culture volumes, in the range of 500–1000 ml, are therefore needed if substantial quantities of λ DNA are to be obtained.

3.3.1   Growth of cultures to obtain a high λ titre

Growing a large volume culture is no problem (bacterial cultures of 50 ℓ and over are common in biotechnology), but obtaining the maximum phage titre requires a certain amount of skill. The naturally occurring λ phage is lysogenic (p. 20), and an infected culture consists mainly of cells carrying the prophage integrated into the bacterial DNA (Figure 2.7). The extracellular λ titre is extremely low under these circumstances.

To get a high yield of extracellular λ, the culture must be **induced**, so that all the bacteria enter the lytic phase of the infection cycle, resulting in cell death and release of λ particles into the medium. Induction is normally very difficult to control, but most laboratory strains of λ carry a **temperature-sensitive (ts) mutation** in the *cI* gene. This is one of the genes that are responsible for maintaining the phage in the integrated state. If inactivated by a mutation, the *cI* gene no longer functions correctly and the switch to lysis occurs.

In the *cI*ts mutation, the *cI* gene is functional at 30°C, at which temperature normal lysogeny can occur. But at 42°C, the *cI*ts gene product does not work properly, and lysogeny cannot be maintained. A culture of *E. coli* infected with λ *cI*ts can therefore be induced to produce extracellular phage by transferring from 30°C to 42°C (Figure 3.17).

3.3.2   Preparation of non-lysogenic λ phage

Although most λ strains are lysogenic, many cloning vectors derived from λ are modified, by deletions of the *cI* and other genes, so that lysogeny never occurs. These phages cannot integrate into the bacterial genome and can infect cells only by a lytic cycle (p. 20).

With these phages the key to obtaining a high titre lies in the way in which the culture is grown, in particular the stage at which the cells are infected by adding phage particles. If phages are added before the cells are dividing at their maximal rate, then all the cells are lysed very quickly, resulting in a low titre (Figure 3.18(a)). On the other hand, if the cell density is too high when the phages are added, then the culture will never be completely lysed, and again the phage titre will be low (Figure 3.18(b)). The ideal situation is when the age of the culture, and the size of the phage inoculum, are balanced such

Figure 3.17 Induction of a λ *cIts* lysogen by transferring from 30°C to 42°C.

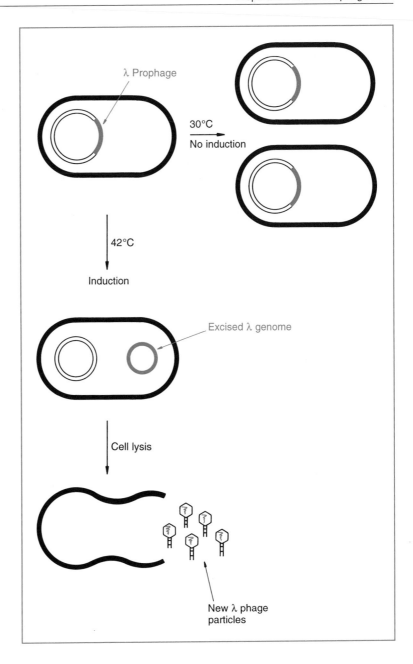

λ Prophage

30°C
No induction

42°C
Induction

Excised λ genome

Cell lysis

New λ phage particles

that the culture continues to grow, but eventually all the cells are infected and lysed (Figure 3.18(c)). As can be imagined, skill and experience are needed to judge the matter to perfection.

3.3.3 Collection of phage from an infected culture

The remains of lysed bacterial cells, along with any intact cells that are inadvertently left over, can be removed from an infected culture by

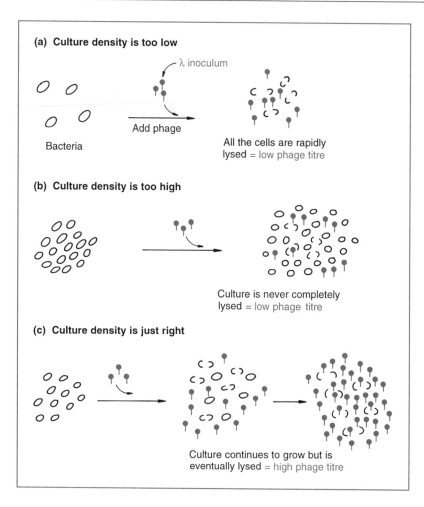

Figure 3.18 Achieving the right balance between culture age and inoculum size when preparing a sample of a non-lysogenic phage.

(a) Culture density is too low

λ inoculum

Bacteria

Add phage

All the cells are rapidly lysed = low phage titre

(b) Culture density is too high

Culture is never completely lysed = low phage titre

(c) Culture density is just right

Culture continues to grow but is eventually lysed = high phage titre

centrifugation, leaving the phage particles in suspension (Figure 3.16). The problem now is to reduce the size of the suspension to 5 ml or less, a manageable size for DNA extraction.

Phage particles are so small that they are pelleted only by very high speed centrifugation. Collection of phages is therefore usually achieved by precipitation with **polyethylene glycol (PEG)**. This is a long-chain polymeric compound which, in the presence of salt, absorbs water, thereby causing macromolecular assemblies such as phage particles to precipitate. The precipitate can then be collected by centrifugation, and redissolved in a suitably small volume (Figure 3.19).

3.3.4   Purification of DNA from λ phage particles

Deproteinization of the redissolved PEG precipitate is sometimes sufficient to extract pure phage DNA, but usually λ phages are subjected to an intermediate purification step. This is necessary because the PEG precipitate also contains a certain amount of bacterial debris, possibly including unwanted

Figure 3.19 Collection of phage particles by polyethylene glycol (PEG) precipitation.

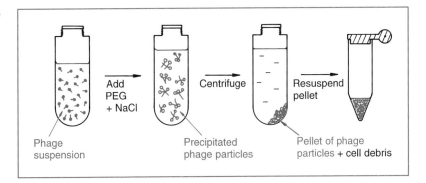

Phage suspension → Add PEG + NaCl → Precipitated phage particles → Centrifuge → Pellet of phage particles + cell debris → Resuspend pellet

Figure 3.20 Purification of λ phage particles by CsCl density gradient centrifugation.

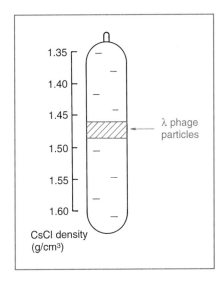

1.35
1.40
1.45
1.50
1.55
1.60

λ phage particles

CsCl density (g/cm³)

cellular DNA. These contaminants can be separated from the λ particles by CsCl density gradient centrifugation. λ particles band in a CsCl gradient at 1.45–1.50 g/cm³ (Figure 3.20), and can be withdrawn from the gradient as described previously for DNA bands (p. 43 and Figure 3.14). Removal of CsCl by dialysis leaves a pure phage preparation from which the DNA can be extracted by either phenol or protease treatment to digest the phage protein coat.

3.3.5   Purification of M13 DNA causes few problems

Most of the differences between the M13 and λ infection cycles are to the advantage of the molecular biologist wishing to prepare M13 DNA. First, the double-stranded replicative form of M13 (p. 25), which behaves like a high copy number plasmid, is very easily purified by the standard procedures for plasmid preparation. A cell extract is prepared from cells infected with M13, and the replicative form separated from bacterial DNA by, for example, EtBr–CsCl density gradient centrifugation.

However, the single-stranded form of the M13 genome, contained in the extracellular phage particles, is frequently required. In this respect the big advantage compared with λ is that high titres of M13 are very easy to obtain. As infected cells continually secrete M13 particles into the medium (Figure 2.8), with lysis never occurring, a high M13 titre is achieved simply by growing the infected culture to a high cell density. In fact titres of 10^{12} per ml and above are quite easy to obtain without any special tricks being used. Such high titres mean that significant amounts of single-stranded M13 DNA can be prepared from cultures of small volume – 5 ml or less. Furthermore, as the infected cells are not lysed, there is no problem with cell debris contaminating the phage suspension. Consequently the CsCl density gradient centrifugation step, needed for λ phage preparation, is rarely required with M13.

In summary, single-stranded M13 DNA preparation involves growth of a small volume of infected culture, centrifugation to pellet the bacteria, precipitation of the phage particles with PEG, phenol extraction to remove the phage protein coats, and ethanol precipitation to concentrate the resulting DNA (Figure 3.21).

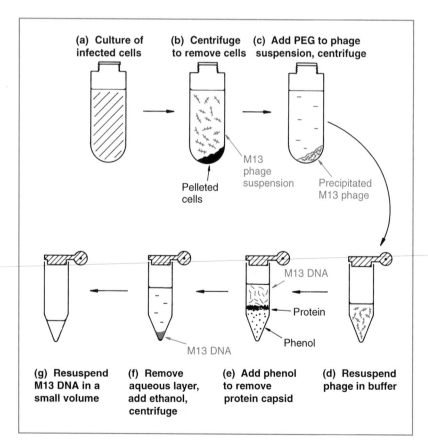

(a) Culture of infected cells

(b) Centrifuge to remove cells

(c) Add PEG to phage suspension, centrifuge

M13 phage suspension

Pelleted cells

Precipitated M13 phage

M13 DNA

Protein

Phenol

M13 DNA

(g) Resuspend M13 DNA in a small volume

(f) Remove aqueous layer, add ethanol, centrifuge

(e) Add phenol to remove protein capsid

(d) Resuspend phage in buffer

Figure 3.21 Preparation of M13 DNA from an infected culture of bacteria.

Further reading

Birnboim, H.C. & Doly, J. (1979) A rapid alkaline extraction procedure for screening recombinant plasmid DNA. *Nucleic Acids Research*, **7**, 1513–23. [A method for preparing plasmid DNA.]

Boom, R., Sol, C.J.A., Salimans, M.M.M., Jansen, C.L., Wertheim van Dillen, P. M. E. & van der Noordaa, J. (1990) Rapid and simple method for purification of nucleic acids. *Journal of Clinical Microbiology*, **28**, 495–503. [The guanidinium thiocyanate and silica method for DNA purification.]

Clewell, D.B. (1972) Nature of ColE1 plasmid replication in *Escherichia coli* in the presence of chloramphenicol. *Journal of Bacteriology*, **110**, 667–76. [The biological basis to plasmid amplification.]

Marmur, J. (1961) A procedure for the isolation of deoxyribonucleic acid from microorganisms. *Journal of Molecular Biology*, **3**, 208–18. [Total cell DNA preparation.]

Radloff, R., Bauer, W. & Vinograd, J. (1967) A dye-buoyant-density method for the detection and isolation of closed-circular duplex DNA. *Proceedings of the National Academy of Sciences of the USA*, **57**, 1514–21. [The original description of ethidium bromide density gradient centrifugation.]

Rogers, S.O. & Bendich, A.J. (1985) Extraction of DNA from milligram amounts of fresh, herbarium and mummified plant tissues. *Plant Molecular Biology*, **5**, 69–76. [The CTAB method.]

Yamamoto, K.R., Alberts, B.M., Benzinger, R., Lawhorne, L. & Trieber, G. (1970) Rapid bacteriophage sedimentation in the presence of polyethylene glycol and its application to large scale virus preparation. *Virology*, **40**, 734–44. [Preparation of λ DNA.]

Zinder, N.D. & Boeke, J.D. (1982) The filamentous phage (Ff) as vectors for recombinant DNA. *Gene*, **19**, 1–10. [Methods for phage growth and DNA preparation.]

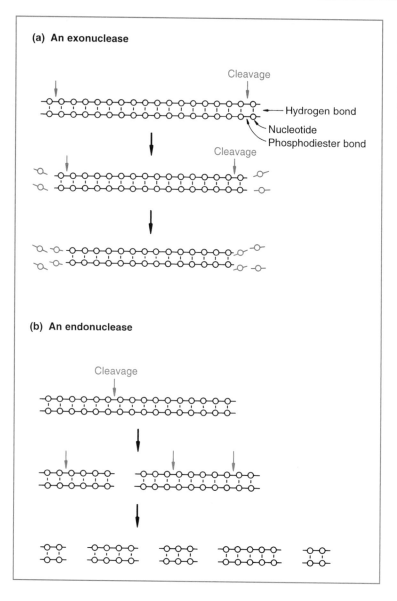

(a) **An exonuclease**

Cleavage

Hydrogen bond

Nucleotide

Phosphodiester bond

Cleavage

(b) **An endonuclease**

Cleavage

Figure 4.1 The reactions catalysed by the two different kinds of nuclease. (a) An exonulease, which removes nucleotides from the end of a DNA molecule. (b) An endonuclease, which breaks internal phosphodiester bonds.

espejiana) is an example of an exonuclease that removes nucleotides from both strands of a double-stranded molecule (Figure 4.2(a)). The greater the length of time that Bal31 is allowed to act on a group of DNA molecules, the shorter the resulting DNA fragments will be. In contrast, enzymes such as *E. coli* exonuclease III degrade just one strand of a double-stranded molecule, leaving single-stranded DNA as the product (Figure 4.2(b)).

The same criterion can be used to classify endonucleases. S1 endonuclease (from the fungus *Aspergillus oryzae*) only cleaves single strands (Figure 4.3(a)), whereas deoxyribonuclease I (DNase I), which is prepared from cow pancreas, cuts both single and double-stranded molecules (Figure 4.3(b)).

Figure 4.2 The reactions catalysed by different types of exonuclease. (a) Bal31, which removes nucleotides from both strands of a double-stranded molecule. (b) Exonuclease III, which removes nucleotides only from the 3′ terminus (see p. 81 for a description of the differences between the 3′ and 5′ termini of a polynucleotide).

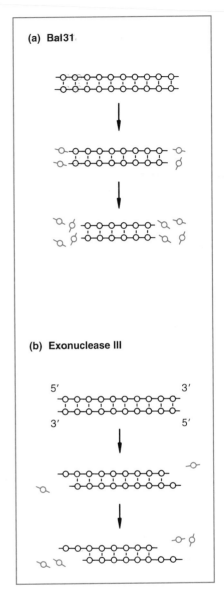

DNase I is non-specific in that it attacks DNA at any internal phosphodiester bond; the end result of prolonged DNase I action is therefore a mixture of mononucleotides and very short oligonucleotides. On the other hand, the special group of enzymes called restriction endonucleases cleave double-stranded DNA only at a limited number of specific recognition sites (Figure 4.3(c)). These important enzymes are described in detail on p. 60.

4.1.2 Ligases

In the cell the function of DNA ligase is to repair single-stranded breaks ('discontinuities') that arise in double-stranded DNA molecules during, for

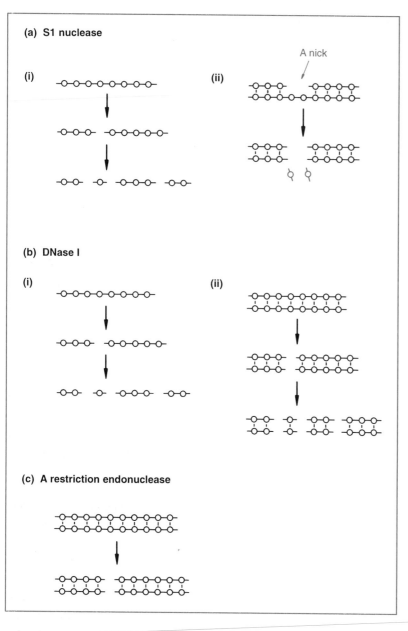

Figure 4.3 The reactions catalysed by different types of endonuclease. (a) S1 nuclease, which cleaves only single-stranded DNA, including single-stranded nicks in mainly double-stranded molecules. (b) DNase I, which cleaves both single- and double-stranded DNA. (c) A restriction endonuclease, which cleaves double-stranded DNA, but only at a limited number of sites.

example, DNA replication. DNA ligases from most organisms can also join together two individual fragments of double-stranded DNA (Figure 4.4). The role of these enzymes in construction of recombinant DNA molecules is described on p. 74.

4.1.3 Polymerases

DNA polymerases are enzymes that synthesize a new strand of DNA complementary to an existing DNA or RNA template (Figure 4.5(a)). Most

Figure 4.4 The two reactions catalysed by DNA ligase. (a) Repair of a discontinuity – a missing phosphodiester bond in one strand of a double-stranded molecule. (b) Joining two molecules together.

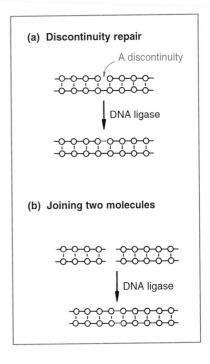

polymerases can function only if the **template** possesses a double-stranded region that acts as a **primer** for initiation of polymerization.

Four types of DNA polymerase are used routinely in genetic engineering. The first is DNA polymerase I, which is usually prepared from *E. coli*. This enzyme attaches to a short single-stranded region (or **nick**) in a mainly double-stranded DNA molecule, and then synthesizes a completely new strand, degrading the existing strand as it proceeds (Figure 4.5(b)). DNA polymerase I is therefore an example of an enzyme with a dual activity – DNA polymerization and DNA degradation.

In fact the polymerase and nuclease activities of DNA polymerase I are controlled by different parts of the enzyme molecule. The nuclease activity is contained in the first 323 amino acids of the polypeptide, so removal of this segment leaves a modified enzyme that retains the polymerase function but is unable to degrade DNA. This modified enzyme, called the **Klenow fragment**, can still synthesize a complementary DNA strand on a single-stranded template, but as it has no nuclease activity it cannot continue the synthesis once the nick is filled in (Figure 4.5(c)). Several other enzymes – natural polymerases and modified versions – have similar properties to the Klenow fragment. The major application of the Klenow fragment and these related polymerases is in DNA sequencing (p. 204).

The *Taq* DNA polymerase used in the polymerase chain reaction (PCR) (Figure 1.2) is the DNA polymerase I enzyme of the bacterium *Thermus aquaticus*. This organism lives in hot springs, and many of its enzymes,

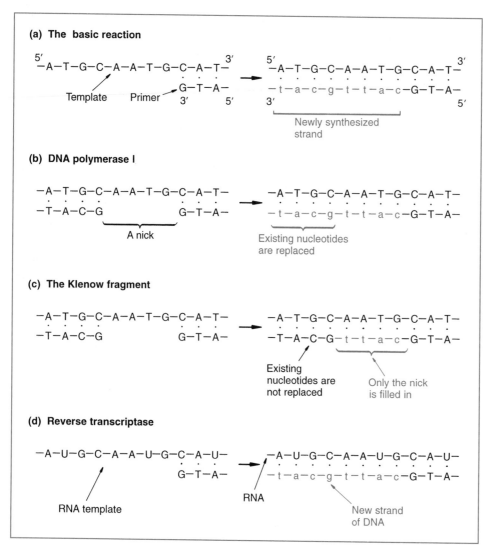

Figure 4.5 The reactions catalysed by DNA polymerases. (a) The basic reaction: a new DNA strand is synthesized in the 5′ to 3′ direction. (b) DNA polymerase I, which initially fills in nicks but then continues to synthesize a new strand, degrading the existing one as it proceeds. (c) The Klenow fragment, which only fills in nicks. (d) Reverse transcriptase, which uses a template of RNA.

including the *Taq* DNA polymerase, are thermostable, meaning that they are resistant to denaturation by heat treatment. This is the special feature of *Taq* DNA polymerase that makes it suitable for PCR, because if it was not thermostable it would be inactivated when the temperature of the reaction is raised to 94°C to denature the DNA.

The final type of DNA polymerase that is important in genetic engineering is **reverse transcriptase**, an enzyme involved in the replication of several kinds of virus. Reverse transcriptase is unique in that it uses as a template not DNA but RNA (Figure 4.5(d)). The ability of this enzyme to synthesize a DNA strand complementary to an RNA template is central to the technique called complementary DNA (cDNA) cloning (p. 164).

4.1.4 DNA modifying enzymes

There are numerous enzymes that modify DNA molecules by addition or removal of specific chemical groups. The most important are as follows.

(1) **Alkaline phosphatase** (from *E. coli*, calf intestinal tissue or arctic shrimp), which removes the phosphate group present at the **5′ terminus** of a DNA molecule (Figure 4.6(a)).

(2) **Polynucleotide kinase** (from *E. coli* infected with T4 phage), which has the reverse effect to alkaline phosphatase, adding phosphate groups onto free 5′ termini (Figure 4.6(b)).

(3) **Terminal deoxynucleotidyl transferase** (from calf thymus tissue), which adds one or more deoxyribonucleotides onto the **3′ terminus** of a DNA molecule (Figure 4.6(c)).

4.1.5 Topoisomerases

The final class of DNA manipulative enzymes are the topoisomerases, which are able to change the conformation of covalently closed-circular DNA (e.g.

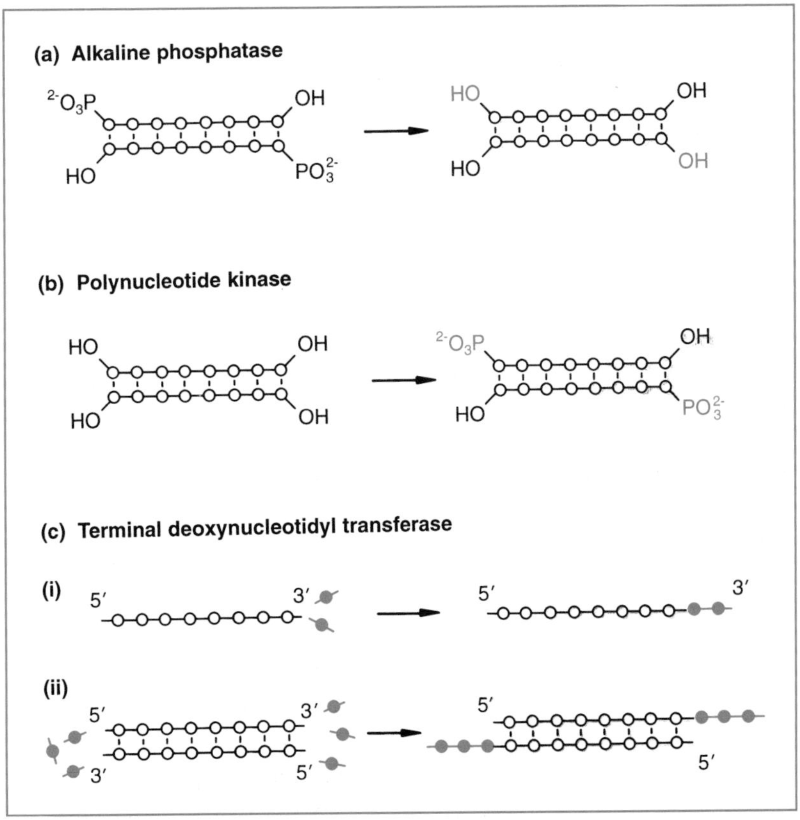

Figure 4.6 The reactions catalysed by DNA modifying enzymes. (a) Alkaline phosphatase, which removes 5′-phosphate groups. (b) Polynucleotide kinase, which attaches 5′-phosphate groups. (c) Terminal deoxynucleotidyl transferase, which attaches deoxyribonucleotides to the 3′ termini of polynucleotides in either (i) single-stranded or (ii) double-stranded molecules.

plasmid DNA molecules) by introducing or removing supercoils (p. 39). Although important in the study of DNA replication, topoisomerases have yet to find a real use in genetic engineering.

4.2 Enzymes for cutting DNA – restriction endonucleases

Gene cloning requires that DNA molecules be cut in a very precise and reproducible fashion. This is illustrated by the way in which the vector is cut during construction of a recombinant DNA molecule (Figure 4.7(a)). Each vector molecule must be cleaved at a single position, to open up the circle so that new DNA can be inserted; a molecule that is cut more than once will be broken into two or more separate fragments and will be of no use as a cloning vehicle. Furthermore, each vector molecule must be cut at exactly the same position on the circle – as will become apparent in later chapters, random cleavage is not satisfactory. It should be clear that a very special type of nuclease is needed to carry out this manipulation.

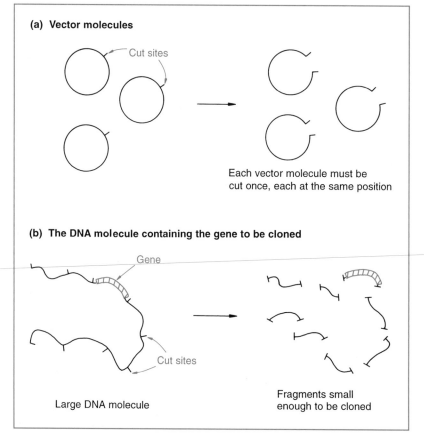

(a) Vector molecules

Cut sites

Each vector molecule must be cut once, each at the same position

(b) The DNA molecule containing the gene to be cloned

Gene

Cut sites

Large DNA molecule

Fragments small enough to be cloned

Figure 4.7 The need for very precise cutting manipulations in a gene cloning experiment.

Often it is also necessary to cleave the DNA that is to be cloned (Figure 4.7(b)). There are two reasons for this. First, if the aim is to clone a single gene, which may consist of only 2 or 3 kb of DNA, that gene will have to be cut out of the large (often greater than 80 kb) DNA molecules produced by skilful use of the preparative techniques described in Chapter 3. Second, large DNA molecules may have to be broken down simply to produce fragments small enough to be carried by the vector. Most cloning vectors exhibit a preference for DNA fragments that fall into a particular size range; M13-based vectors, for example, are very inefficient at cloning DNA molecules of more than 3 kb in length.

Purified restriction endonucleases allow the molecular biologist to cut DNA molecules in the precise, reproducible manner required for gene cloning. The discovery of these enzymes, which led to Nobel Prizes for W. Arber, H. Smith and D. Nathans in 1978, was one of the key breakthroughs in the development of genetic engineering.

4.2.1 The discovery and function of restriction endonucleases

The initial observation that led to the eventual discovery of restriction endonucleases was made in the early 1950s, when it was shown that some strains of bacteria are immune to bacteriophage infection, a phenomenon referred to as **host-controlled restriction**.

The mechanism of restriction is not very complicated, even though it took over 20 years to be fully understood. Restriction occurs because the bacterium produces an enzyme that degrades the phage DNA before it has time to replicate and direct synthesis of new phage particles (Figure 4.8(a)). The bacterium's own DNA, the destruction of which would of course be lethal, is protected from attack because it carries additional methyl groups that block the degradative enzyme action (Figure 4.8(b)).

These degradative enzymes are called restriction endonucleases and are synthesized by many, perhaps all, species of bacteria; over 1200 different ones have now been characterized. Three different classes of restriction endonuclease are recognized, each distinguished by a slightly different mode of action. Types I and III are rather complex and have only a limited role in genetic engineering. Type II restriction endonucleases, on the other hand, are the cutting enzymes that are so important in gene cloning.

4.2.2 Type II restriction endonucleases cut DNA at specific nucleotide sequences

The central feature of type II restriction endonucleases (which will be referred to simply as 'restriction endonucleases' from now on) is that each enzyme has a specific recognition sequence at which it cuts a DNA molecule. A particular enzyme cleaves DNA at the recognition sequence and nowhere

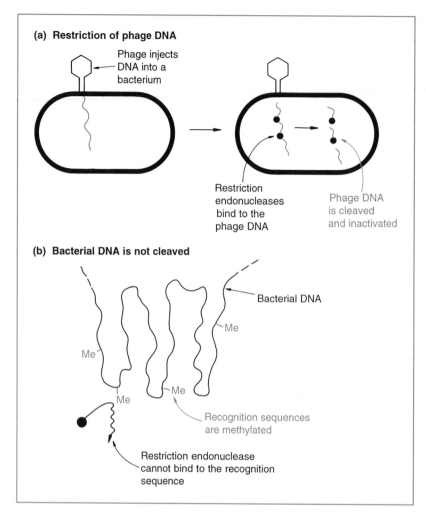

(a) Restriction of phage DNA

Phage injects
DNA into a
bacterium

Restriction
endonucleases
bind to the
phage DNA

Phage DNA
is cleaved
and inactivated

(b) Bacterial DNA is not cleaved

Bacterial DNA

Me

Me

Me

Me

Recognition sequences
are methylated

Restriction endonuclease
cannot bind to the recognition
sequence

Figure 4.8 The function of a restriction endonuclease in a bacterial cell: (a) phage DNA is cleaved, but (b) bacterial DNA is not.

else. For example, the restriction endonuclease called *Pvu*I (isolated from *Proteus vulgaris*) cuts DNA only at the hexanucleotide CGATCG. In contrast, a second enzyme from the same bacterium, called *Pvu*II, cuts at a different hexanucleotide, in this case CAGCTG.

Many restriction endonucleases recognize hexanucleotide target sites, but others cut at four, five or even eight nucleotide sequences. *Sau*3A (from *Staphylococcus aureus* strain 3A) recognizes GATC, and *Alu*I (*Arthrobacter luteus*) cuts at AGCT. There are also examples of restriction endonucleases with degenerate recognition sequences, meaning that they cut DNA at any one of a family of related sites. *Hinf*I (*Haemophilus influenzae* strain R_f), for instance, recognizes GANTC, so cuts at GAATC, GATTC, GAGTC and GACTC.

The restriction sequences for some of the most frequently used restriction endonucleases are listed in Table 4.1.

Table 4.1 The recognition sequences for some of the most frequently used restriction endonucleases.

Enzyme	Organism	Recognition sequence[a]	Blunt or sticky end
*Eco*RI	*Escherichia coli*	GAATTC	Sticky
*Bam*HI	*Bacillus amyloliquefaciens*	GGATCC	Sticky
*Bgl*II	*Bacillus globigii*	AGATCT	Sticky
*Pvu*I	*Proteus vulgaris*	CGATCG	Sticky
*Pvu*II	*Proteus vulgaris*	CAGCTG	Blunt
*Hind*III	*Haemophilus influenzae* R_d	AAGCTT	Sticky
*Hinf*I	*Haemophilus influenzae* R_f	GANTC	Sticky
*Sau*3A	*Staphylococcus aureus*	GATC	Sticky
*Alu*I	*Arthrobacter luteus*	AGCT	Blunt
*Taq*I	*Thermus aquaticus*	TCGA	Sticky
*Hae*III	*Haemophilus aegyptius*	GGCC	Blunt
*Not*I	*Nocardia otitidis-caviarum*	GCGGCCGC	Sticky
*Sfi*I	*Streptomyces fimbriatus*	GGCCNNNNNGGCC	Sticky

[a] The sequence shown is that of one strand, given in the 5′ to 3′ direction. Note that almost all recognition sequences are palindromes: when both strands are considered they read the same in each direction, for example:

$$5'-G\ A\ A\ T\ T\ C-3'$$

*Eco*RI

$$3'-C\ T\ T\ A\ A\ G-5'$$

4.2.3

Blunt ends and sticky ends

The exact nature of the cut produced by a restriction endonuclease is of considerable importance in the design of a gene cloning experiment. Many restriction endonucleases make a simple double-stranded cut in the middle of the recognition sequence (Figure 4.9(a)), resulting in a **blunt end** or **flush end**. *Pvu*II and *Alu*I are examples of blunt end cutters.

However, quite a large number of restriction endonucleases cut DNA in a slightly different way. With these enzymes the two DNA strands are not cut at exactly the same position. Instead the cleavage is staggered, usually by two or four nucleotides, so that the resulting DNA fragments have short single-stranded overhangs at each end (Figure 4.9(b)). These are called sticky or cohesive ends, as base pairing between them can stick the DNA molecule back together again (recall that sticky ends were encountered on p. 23 during the description of λ phage replication). One important feature of sticky end enzymes is that restriction endonucleases with different recognition sequences may produce the same sticky ends. *Bam*HI (recognition sequence GGATCC) and *Bgl*II (AGATCT) are examples – both produce GATC sticky ends (Figure 4.9(c)). The same sticky end is also produced by *Sau*3A, which recognizes only the tetranucleotide GATC. Fragments of DNA

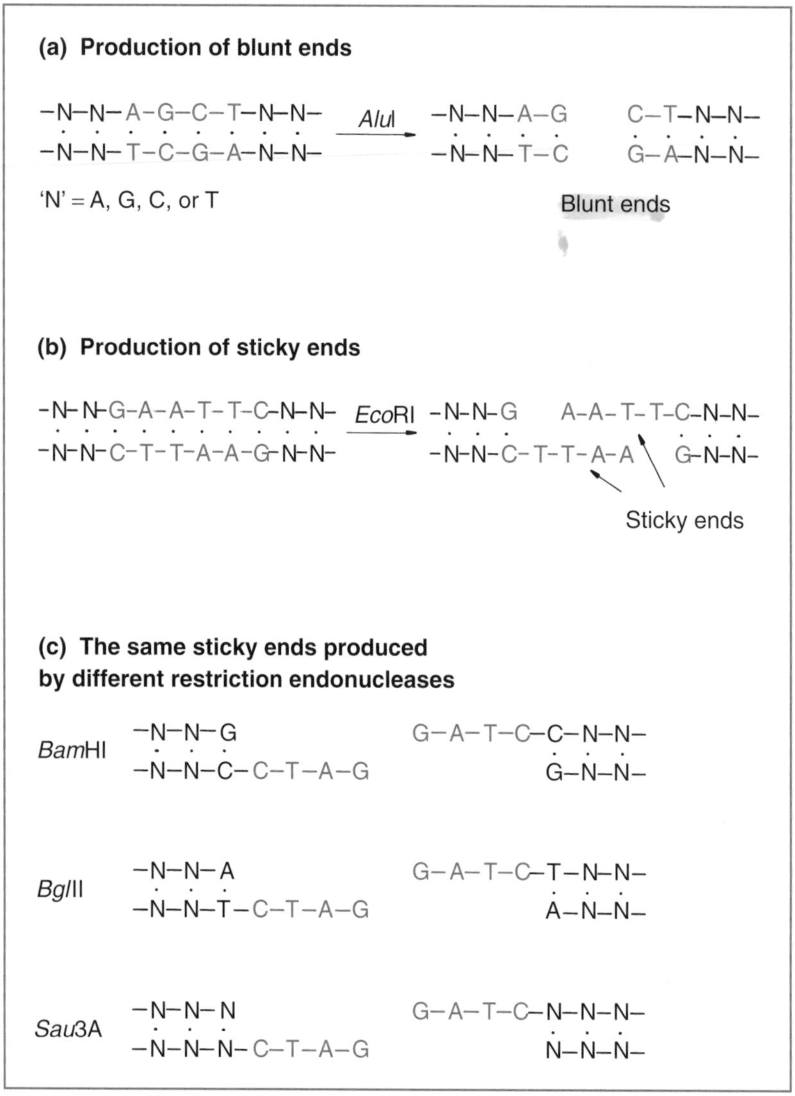

(a) Production of blunt ends

```
−N−N−A−G−C−T−N−N−   AluI    −N−N−A−G       C−T−N−N−
 · · · · · · · ·             · · · ·        · · · ·
−N−N−T−C−G−A−N−N−            −N−N−T−C       G−A−N−N−
```

'N' = A, G, C, or T   Blunt ends

(b) Production of sticky ends

```
−N−N−G−A−A−T−T−C−N−N−  EcoRI  −N−N−G     A−A−T−T−C−N−N−
 · · · · · · · · · ·          · · ·      · · · ·
−N−N−C−T−T−A−A−G−N−N−         −N−N−C−T−T−A−A   G−N−N−
```

Sticky ends

(c) The same sticky ends produced by different restriction endonucleases

BamHI
```
      −N−N−G             G−A−T−C−C−N−N−
       · ·                 · ·
  −N−N−C−C−T−A−G              G−N−N−
```

BglII
```
      −N−N−A             G−A−T−C−T−N−N−
       · ·                 · ·
  −N−N−T−C−T−A−G            A−N−N−
```

Sau3A
```
      −N−N−N             G−A−T−C−N−N−N−
       · ·                 · ·
  −N−N−N−C−T−A−G          N−N−N−
```

Figure 4.9 The ends produced by cleavage of DNA with different restriction endonucleases. (a) A blunt end produced by AluI. (b) A sticky end produced by EcoRI. (c) The same sticky ends produced by BamHI, BglII and Sau3A.

produced by cleavage with either of these enzymes can be joined to each other, as each fragment carries a complementary sticky end.

4.2.4   The frequency of recognition sequences in a DNA molecule

The number of recognition sequences for a particular restriction endonuclease in a DNA molecule of known length can be calculated mathematically. A tetranucleotide sequence (e.g. GATC) should occur once every $4^4 = 256$ nucleotides, and a hexanucleotide (e.g. GGATCC) once every $4^6 = 4096$ nucleotides. These calculations assume that the nucleotides are ordered in a random fashion and that the four different nucleotides are present in equal proportions (i.e. the GC content = 50%). In practice, neither of these assumptions is entirely valid. For example, the λ DNA molecule, at 49 kb, should contain about 12 sites for a restriction endonuclease with a hexanucleotide recognition sequence. In fact these recognition sites occur less frequently (e.g. six for *Bgl*II, five for *Bam*HI and only two for *Sal*I), a reflection of the fact that the GC content for λ is rather less than 50% (Figure 4.10(a)).

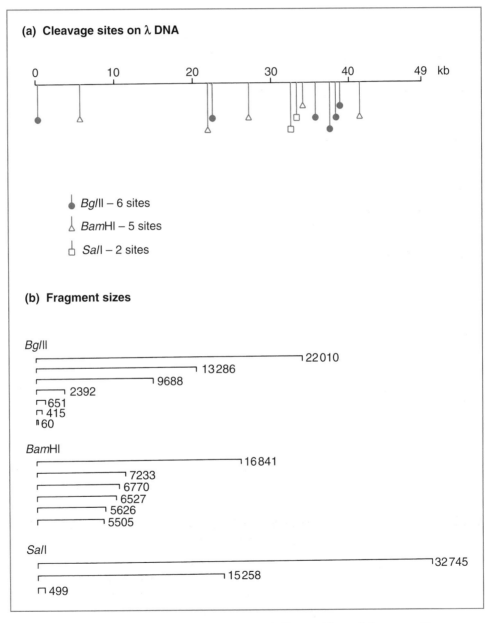

Figure 4.10 Restriction of the λ DNA molecule. (a) The positions of the recognition sequences for *Bgl*II, *Bam*HI and *Sal*I. (b) The fragments produced by cleavage with each of these restriction endonucleases; the numbers are the fragment sizes in base pairs.

Furthermore, restriction sites are generally not evenly spaced out along a DNA molecule. If they were then digestion with a particular restriction endonuclease would give fragments of roughly equal sizes. Figure 4.10(b) shows the fragments produced by cutting λ DNA with BglII, BamHI and SalI. In each case there is a considerable spread of fragment sizes, indicating that in λ DNA the nucleotides are not randomly ordered.

The lesson to be learned from Figure 4.10 is that although mathematics may give an idea of how many restriction sites to expect in a given DNA molecule, only experimental analysis can provide the true picture. We must therefore move on to consider how restriction endonucleases are used in the laboratory.

4.2.5 Performing a restriction digest in the laboratory

As an example, we will consider how to digest a sample of λ DNA (concentration 125 μg/ml) with BglII.

First of all the required amount of DNA must be pipetted into a test tube. The amount of DNA that will be restricted depends on the nature of the experiment; in this case we shall digest 2 μg of λ DNA, which is contained in 16 μl of the sample (Figure 4.11(a)). Very accurate micropipettes will therefore be needed.

Clearly the other main component in the reaction will be the restriction endonuclease, obtained from a commercial supplier as a pure solution of known concentration. But before adding the enzyme, the solution containing the DNA must be adjusted to provide the correct conditions to ensure maximal activity of the enzyme. Most restriction endonucleases function adequately at pH 7.4, but different enzymes vary in their requirements for ionic strength (usually provided by sodium chloride (NaCl)) and magnesium (Mg^{2+}) concentration (all type II restriction endonucleases require Mg^{2+} in order to function). It is also advisable to add a reducing agent, such as dithiothreitol (DTT), which stabilizes the enzyme and prevents its inactivation. Providing the right conditions for the enzyme is very important – incorrect NaCl or Mg^{2+} concentrations may not only decrease the activity of the restriction endonuclease, but may also cause changes in the specificity of the enzyme, so that DNA cleavage occurs at additional, non-standard recognition sequences.

The composition of a suitable buffer for BglII is shown in Table 4.2. This buffer is ten times the working concentration, and is diluted by being added to the reaction mixture. In our example, a suitable final volume for the reaction mixture would be 20 μl, so we add 2 μl of 10 × BglII buffer to the 16 μl of DNA already present (Figure 4.11(b)).

The restriction endonuclease can now be added. By convention, 1 unit of enzyme is defined as the quantity needed to cut 1 μg of DNA in 1 hour, so we need 2 units of BglII to cut 2 μg of λ DNA. BglII is frequently obtained at a concentration of 4 units/μl, so 0.5 μl is sufficient to cleave the DNA. The

Figure 4.11 Performing a restriction digest in the laboratory (see text for details).

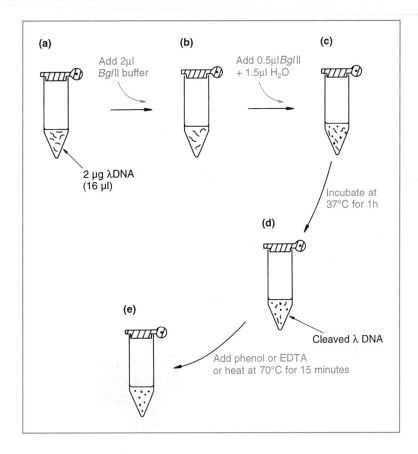

Table 4.2 A 10× buffer suitable for restriction of DNA with *Bgl*II.

Component	Concentration (mM)
Tris–HCl, pH 7.4	500
MgCl$_2$	100
NaCl	500
Dithiothreitol	10

final ingredients in the reaction mixture are therefore 0.5 µl *Bgl*II + 1.5 µl water, giving a final volume of 20 µl (Figure 4.11(c)).

The last factor to consider is incubation temperature. Most restriction endonucleases, including *Bgl*II, work best at 37°C, but a few have different requirements. *Taq*I, for example, is a restriction enzyme from *Thermus aquaticus* and, like *Taq* DNA polymerase, has a high working temperature. Restriction digests with *Taq*I must be incubated at 65°C to obtain maximum enzyme activity.

After 1 hour the restriction should be complete (Figure 4.11(d)). If the DNA fragments produced by restriction are to be used in cloning experiments, the enzyme must somehow be destroyed so that it does not accidentally digest other DNA molecules that may be added at a later stage. There are several ways of 'killing' the enzyme. For many a short incubation at 70°C is sufficient, for others phenol extraction or the addition of ethylenediamine tetraacetic acid (EDTA) (which binds Mg^{2+} ions preventing restriction endonuclease action) is used (Figure 4.11(e)).

4.2.6 Analysing the result of restriction endonuclease cleavage

A restriction digest results in a number of DNA fragments, the sizes of which depend on the exact positions of the recognition sequences for the endonuclease in the original molecule (Figure 4.10). Clearly a way of determining the number and sizes of the fragments is needed if restriction endonucleases are to be of use in gene cloning. Whether or not a DNA molecule is cut at all can be determined fairly easily by testing the viscosity of the solution. Larger DNA molecules result in a more viscous solution than smaller ones, so cleavage is associated with a decrease in viscosity. However, working out the number and sizes of the individual cleavage products is more difficult. In fact, for several years this was one of the most tedious aspects of experiments involving DNA. Eventually the problems were solved in the early 1970s when the technique of gel electrophoresis was developed.

Separation of molecules by gel electrophoresis

DNA molecules, like proteins and many other biological compounds, carry an electric charge, negative in the case of DNA. Consequently, when DNA molecules are placed in an electric field they migrate towards the positive pole (Figure 4.12 (a)). The rate of migration of a molecule depends on two factors, its shape and its charge-to-mass ratio. Unfortunately, most DNA molecules are the same shape and all have very similar charge-to-mass ratios. Fragments of different sizes cannot therefore be separated by standard **electrophoresis.**

However, the size of the DNA molecule does become a factor if the electrophoresis is performed in a gel. A gel, which is usually made of agarose, polyacrylamide or a mixture of the two, comprises a complex network of pores, through which the DNA molecules must travel to reach the positive electrode. The smaller the DNA molecule, the faster it can migrate through the gel. **Gel electrophoresis** therefore separates DNA molecules according to their size (Figure 4.12(b)).

In practice the composition of the gel determines the sizes of the DNA molecules that can be separated. A 0.5 cm thick slab of 0.5% agarose, which has relatively large pores, would be used for molecules in the size range 1–30 kb, allowing, for example, molecules of 10 and 12 kb to be clearly distinguished. At the other end of the scale, a very thin (0.3 mm) 40%

Figure 4.12 (a) Standard electrophoresis does not separate DNA fragments of different sizes, whereas (b) gel electrophoresis does.

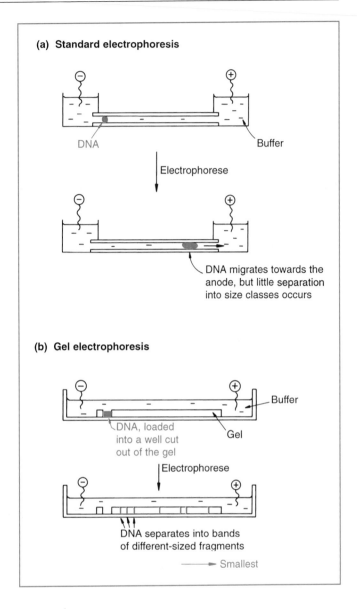

polyacrylamide gel, with extremely small pores, would be used to separate much smaller DNA molecules, in the range 1–300 bp, and could distinguish molecules differing in length by just a single nucleotide.

Visualizing DNA molecules in a gel

Staining

The easiest way to see the results of a gel electrophoresis experiment is to stain the gel with a compound that makes the DNA visible. Ethidium bromide (EtBr), already described on p. 42 as a means of visualizing DNA in caesium chloride (CsCl) gradients, is also routinely used to stain DNA in

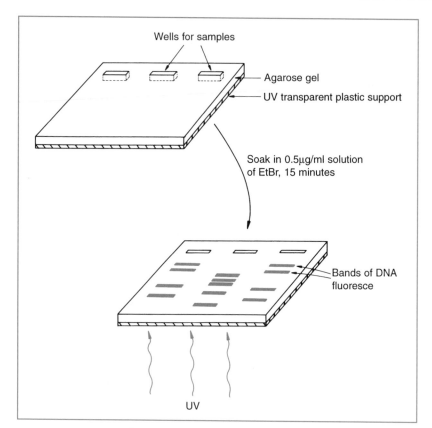

Figure 4.13 Visualizing DNA bands in an agarose gel by EtBr staining and ultraviolet (UV) irradiation.

Wells for samples

Agarose gel

UV transparent plastic support

Soak in 0.5μg/ml solution of EtBr, 15 minutes

Bands of DNA fluoresce

UV

agarose and polyacrylamide gels (Figure 4.13). Bands showing the positions of the different size classes of DNA fragment are clearly visible under ultraviolet irradiation after EtBr staining, so long as sufficient DNA is present. Unfortunately, the procedure is very hazardous because ethidium bromide is a powerful mutagen and the ultraviolet radiation used to visualize the DNA can cause severe burns. For this reason, non-mutagenic dyes that stain DNA green or blue, and which do not require ultraviolet irradiation in order for the results to be seen, are now used in many laboratories.

Autoradiography of radioactively labelled DNA

The one drawback with staining is that there is a limit to its sensitivity. If less than about 10 ng of DNA is present per band, it is unlikely that the bands will show up after staining. For small amounts of DNA a more sensitive detection method is needed.

Autoradiography provides an answer. If the DNA is **labelled** before electrophoresis, by incorporation of a **radioactive marker** into the individual molecules, the DNA can be visualized by placing an X-ray-sensitive photographic film over the gel. The radioactive DNA exposes the film, revealing the banding pattern (Figure 4.14).

Figure 4.14 The use of autoradiography to visualize radioactively labelled DNA in an agarose gel.

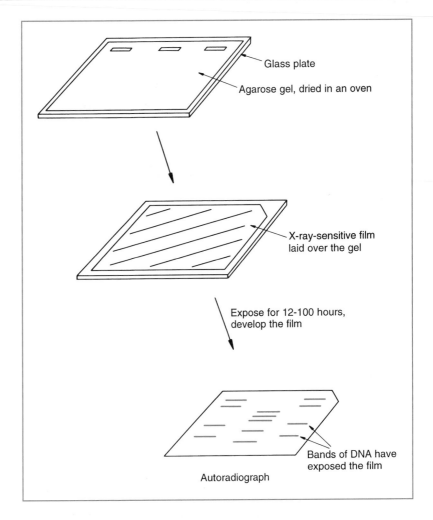

Glass plate

Agarose gel, dried in an oven

X-ray-sensitive film laid over the gel

Expose for 12-100 hours, develop the film

Bands of DNA have exposed the film

Autoradiograph

A DNA molecule is usually labelled by incorporating nucleotides that carry a radioactive isotope of phosphorus, ^{32}P (Figure 4.15(a)). Several methods are available, the most popular being **nick translation** and **end filling**.

Nick translation refers to the activity of DNA polymerase I (p. 57). Most purified samples of DNA contain some nicked molecules, however carefully the preparation has been carried out, which means that DNA polymerase I is able to attach to the DNA and catalyse a strand replacement reaction (Figure 4.5(b)). This reaction requires a supply of nucleotides: if one of these is radioactively labelled, the DNA molecule will itself become labelled (Figure 4.15(b)).

Nick translation can be used to label any DNA molecule but may under some circumstances also cause DNA cleavage. End filling is a gentler method that rarely causes breakage of the DNA, but unfortunately can only be used to label DNA molecules that have sticky ends. The enzyme used is the Klenow fragment (p. 57), which 'fills in' a sticky end by synthesizing the

71

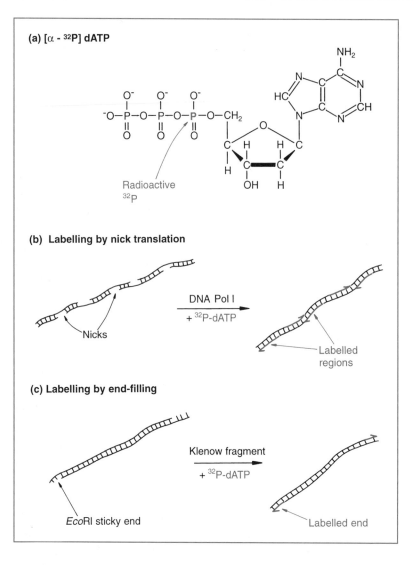

(a) [α - ³²P] dATP

Radioactive
³²P

(b) Labelling by nick translation

Nicks

DNA Pol I
+ ³²P-dATP

Labelled
regions

(c) Labelling by end-filling

EcoRI sticky end

Klenow fragment
+ ³²P-dATP

Labelled end

Figure 4.15 Radioactive labelling: (a) structure of α-³²P-deoxyadenosine triphosphate ([α-³²P]dATP), (b) labelling DNA by nick translation, and (c) labelling DNA by end filling.

complementary strand (Figure 4.15(c)). As with nick translation, if the end filling reaction is carried out in the presence of labelled nucleotides, the DNA itself becomes labelled.

Both nick translation and end filling can label DNA to such an extent that very small quantities can be detected in gels by autoradiography. As little as 2 ng of DNA per band can be visualized under ideal conditions.

4.2.7   Estimation of the sizes of DNA molecules

Gel electrophoresis will separate different sized DNA molecules, with the smallest molecules travelling the greatest distance towards the positive electrode. If several DNA fragments of varying sizes are present (the result of a successful restriction digest, for example), then a series of bands appears in the gel. How can the sizes of these fragments be determined?

The most accurate method is to make use of the mathematical relationship that links migration rate to molecular weight. The relevant formula is:

$$D = a - b(\log M)$$

Where D is the distance moved, M is the molecular weight, and a and b are constants that depend on the electrophoresis conditions.

A much simpler though less precise way of estimating DNA fragment sizes is generally used. A standard restriction digest, comprising fragments of known size, is usually included in each electrophoresis gel that is run. Restriction digests of λ DNA are often used in this way as size markers. For example, *Hind*III cleaves λ DNA into eight fragments, ranging in size from 125 bp for the smallest to over 23 kb for the largest. As the sizes of the fragments in this digest are known, the fragment sizes in the experimental digest can be estimated by comparing the positions of the bands in the two tracks (Figure 4.16). Although not precise, this method can be performed with as little as a 5% error, which is quite satisfactory for most purposes.

4.2.8

Mapping the positions of different restriction sites on a DNA molecule

So far we have considered how to determine the number and sizes of the DNA fragments produced by restriction endonuclease cleavage. The next step in **restriction analysis** is to construct a map showing the relative positions on the DNA molecule of the recognition sequences for a number of different enzymes. Only when a **restriction map** is available can the correct restriction endonucleases be selected for the particular cutting manipulation that is required (Figure 4.17).

To construct a restriction map, a series of restriction digests must be performed. First, the number and sizes of the fragments produced by each restriction endonuclease must be determined by gel electrophoresis, followed by comparison with size markers (Figure 4.18). This information must then be supplemented by a series of **double digestions**, in which the DNA is cut by two restriction endonucleases at once. It may be possible to perform a double digestion in one step, if both enzymes have similar requirements for pH, Mg^{2+} concentration, etc. Alternatively, the two digestions may have to be carried out one after the other, adjusting the reaction mixture after the first digestion to provide a different set of conditions for the second enzyme.

Comparing the results of single and double digests will allow many, if not all, of the restriction sites to be mapped (Figure 4.18). Ambiguities can usually be resolved by **partial digestion**, carried out under conditions that result in cleavage of only a limited number of the restriction sites on any DNA molecule. Partial digestion is usually achieved by reducing the incubation period, so the enzyme does not have time to cut all the restriction sites, or by incubating at a low temperature (e.g. 4°C rather than 37°C), which limits the activity of the enzyme.

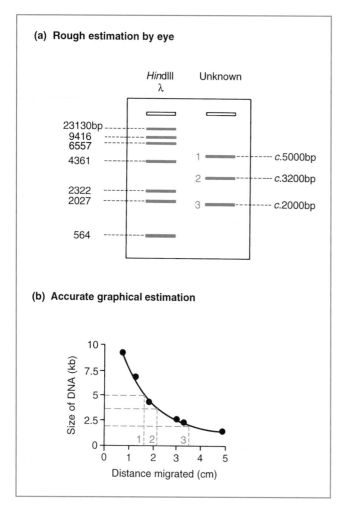

(a) Rough estimation by eye

*Hind*III
λ

Unknown

23130bp
9416
6557
4361
2322
2027
564

1 — *c.*5000bp
2 — *c.*3200bp
3 — *c.*2000bp

(b) Accurate graphical estimation

Size of DNA (kb)

10
7.5
5
2.5
0

1 2   3

0   1   2   3   4   5
Distance migrated (cm)

Figure 4.16 Estimation of the sizes of DNA fragments in an agarose gel. (a) A rough estimate of fragment size can be obtained by eye. (b) A more accurate measurement of fragment size is gained by using the mobilities of the *Hind*III–λ fragments to construct a calibration curve; the sizes of the unknown fragments can then be determined from the distances they have migrated.

The result of a partial digestion is a complex pattern of bands in an electrophoresis gel. As well as the standard fragments, produced by total digestion, additional sizes are seen. These are molecules that comprise two adjacent restriction fragments, separated by a site that has not been cleaved. Their sizes indicate which restriction fragments in the complete digest are next to one another in the uncut molecule (Figure 4.18).

4.3

Ligation – joining DNA molecules together

The final step in construction of a recombinant DNA molecule is the joining together of the vector molecule and the DNA to be cloned (Figure 4.19). This process is referred to as ligation, and the enzyme that catalyses the reaction is called DNA ligase.

Figure 4.17 Using a restriction map to work out which restriction endonucleases should be used to obtain DNA fragments containing individual genes.

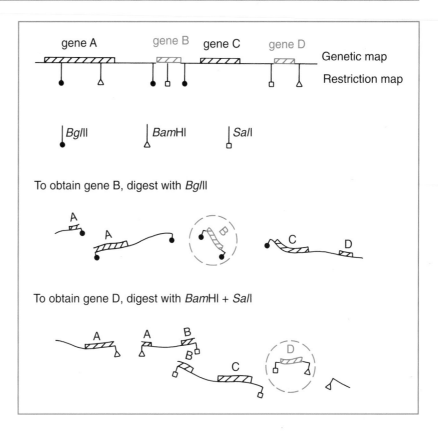

4.3.1 The mode of action of DNA ligase

All living cells produce DNA ligases, but the enzyme used in genetic engineering is usually purified from *E. coli* bacteria that have been infected with T4 phage. Within the cell the enzyme carries out the very important function of repairing any discontinuities that may arise in one of the strands of a double-stranded molecule (Figure 4.4(a)). A discontinuity is quite simply a position where a phosphodiester bond between adjacent nucleotides is missing (contrast this with a nick, where one or more nucleotides are absent). Although discontinuities may arise by chance breakage of the cell's DNA molecules, they are also a natural result of processes such as DNA replication and recombination. Ligases therefore play several vital roles in the cell.

In the test tube, purified DNA ligases, as well as repairing single-strand discontinuities, can also join together individual DNA molecules or the two ends of the same molecule. The chemical reaction involved in ligating two molecules is exactly the same as discontinuity repair, except that two phosphodiester bonds must be made, one for each strand (Figure 4.20(a)).

4.3.2 Sticky ends increase the efficiency of ligation

The ligation reaction in Figure 4.20(a) shows two blunt-ended fragments being joined together. Although this reaction can be carried out in the test

Single and double digestions

Enzyme	Number of fragments	Sizes (kb)
*Xba*l	2	24.0, 24.5
*Xho*l	2	15.0, 33.5
*Kpn*l	3	1.5, 17.0, 30.0
*Xba*l + *Xho*l	3	9.0, 15.0, 24.5
*Xba*l + *Kpn*l	4	1.5, 6.0, 17.0, 24.0

Conclusions:

(1) As λ DNA is linear, the number of restriction sites for each enzyme is *Xba*l 1, *Xho*l 1, *Kpn*l 2.

(2) The *Xba*l and *Xho*l sites can be mapped:

*Xba*l fragments   24.0   24.5

*Xba*l − *Xho*l fragments   9.0   15.0   24.5

*Xho*l fragments   15.0   33.5

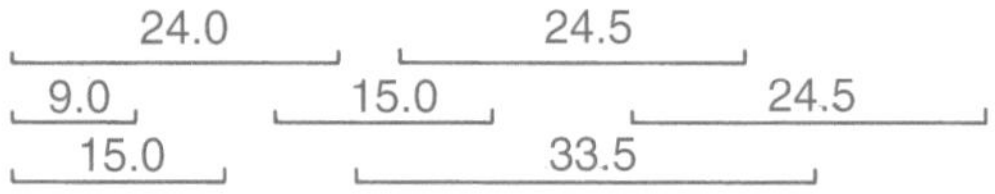

The only possibility is:
    *Xho*l *Xba*l
    15.0   9.0   24.5

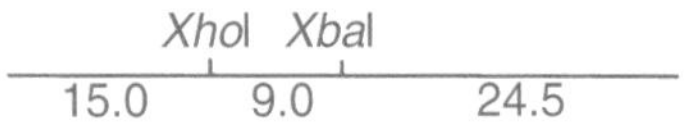

(3) All the *Kpn*l sites fall in the 24.5 kb *Xba*l fragment, as the 24.0 kb fragment is intact after *Xba*l–*Kpn*l double digestion. The order of the *Kpn*l fragments can be determined only by partial digestion.

Partial digestion

Enzyme	Fragment sizes (kb)
*Kpn*l – limiting conditions	1.5, 17.0, 18.5, 30.0, 31.5, 48.5

Conclusions:

(1) 48.5 kb fragment = uncut λ.
(2) 1.5, 17.0 and 30.0 kb fragments are products of complete digestion.
(3) 18.5 and 31.5 kb fragments are products of partial digestion.

The *Kpn*l map must be:
    *Kpn*ls
    30.0 1.5   17.0

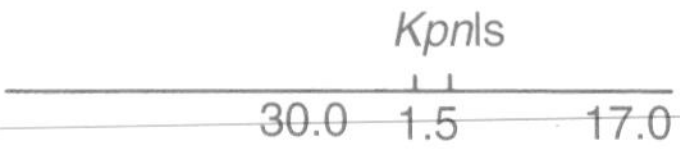

Therefore the complete map is:
    *Xho*l *Xba*l *Kpn*ls
    15.0   9.0 6.0 1.5 17.0

Figure 4.18 Restriction mapping. This example shows how the positions of the *Xba*l, *Xho*l and *Kpn*l sites on the λ DNA molecule can be determined.

Figure 4.19 Ligation: the final step in construction of a recombinant DNA molecule.

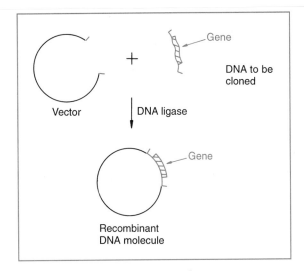

Figure 4.20 The different joining reactions catalysed by DNA ligase: (a) ligation of blunt-ended molecules, and (b) ligation of sticky-ended molecules.

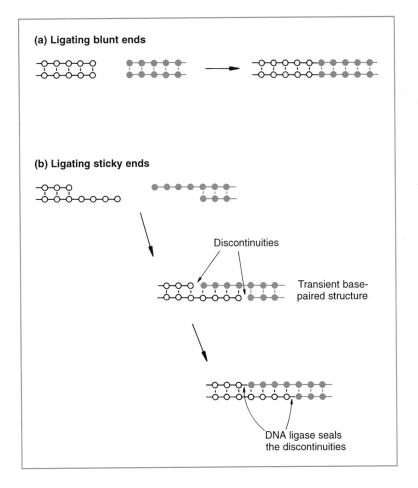

tube, it is not very efficient. This is because the ligase is unable to 'catch hold' of the molecule to be ligated, and has to wait for chance associations to bring the ends together. If possible, blunt end ligation should be performed at high DNA concentrations, to increase the chances of the ends of the molecules coming together in the correct way.

In contrast, ligation of complementary sticky ends is much more efficient. This is because compatible sticky ends can base pair with one another by hydrogen bonding (Figure 4.20(b)), forming a relatively stable structure for the enzyme to work on. If the phosphodiester bonds are not synthesized fairly quickly the sticky ends will fall apart again. These transient, base-paired structures do, however, increase the efficiency of ligation by increasing the length of time the ends are in contact with one another.

4.3.3   Putting sticky ends onto a blunt-ended molecule

For the reasons detailed in the preceding section, compatible sticky ends are desirable on the DNA molecules to be ligated together in a gene cloning experiment. Often these sticky ends can be provided by digesting both the vector and the DNA to be cloned with the same restriction endonuclease, or with different enzymes that produce the same sticky end; however, it is not always possible to do this. A common situation is where the vector molecule has sticky ends, but the DNA fragments to be cloned are blunt-ended. Under these circumstances one of three methods can be used to put the correct sticky ends onto the DNA fragments.

Linkers

The first of these methods involves the use of **linkers**. These are short pieces of double-stranded DNA, of known nucleotide sequence, that are synthesized in the test tube. A typical linker is shown in Figure 4.21(a). It is blunt-ended, but contains a restriction site, *Bam*HI in the example shown. DNA ligase can attach linkers to the ends of larger blunt-ended DNA molecules. Although a blunt end ligation, this particular reaction can be performed very efficiently because synthetic oligonucleotides, such as linkers, can be made in very large amounts and added into the ligation mixture at a high concentration.

More than one linker will attach to each end of the DNA molecule, producing the chain structure shown in Figure 4.21(b). However, digestion with *Bam*HI cleaves the chains at the recognition sequences, producing a large number of cleaved linkers and the original DNA fragment, now carrying *Bam*HI sticky ends. This modified fragment is ready for ligation into a cloning vector restricted with *Bam*HI.

Adaptors

There is one potential drawback with the use of linkers. Consider what would happen if the blunt-ended molecule shown in Figure 4.21(b) contained one

Figure 4.21 Linkers and their use: (a) the structure of a typical linker, and (b) the attachment of linkers to a blunt-ended molecule.

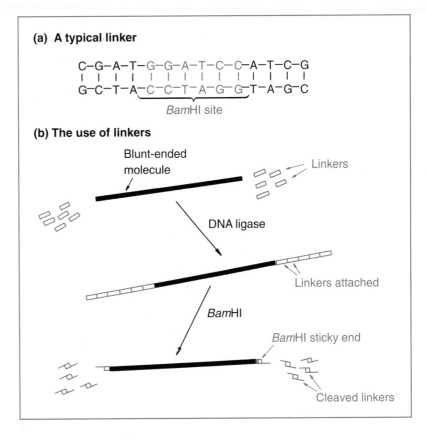

(a) A typical linker

C–G–A–T–G–G–A–T–C–C–A–T–C–G
| | | | | | | | | | | | | |
G–C–T–A–C–C–T–A–G–G–T–A–G–C

*Bam*HI site

(b) The use of linkers

Blunt-ended molecule

Linkers

DNA ligase

Linkers attached

*Bam*HI

*Bam*HI sticky end

Cleaved linkers

or more *Bam*HI recognition sequences. If this was the case the restriction step needed to cleave the linkers and produce the sticky ends would also cleave the blunt-ended molecule (Figure 4.22). The resulting fragments will have the correct sticky ends, but that is no consolation if the gene contained in the blunt-ended fragment has now been broken into pieces.

The second method of attaching sticky ends to a blunt-ended molecule is designed to avoid this problem. **Adaptors**, like linkers, are short synthetic oligonucleotides. But unlike linkers, an adaptor is synthesized so that it already has one sticky end (Figure 4.23(a)). The idea is of course to ligate the blunt end of the adaptor to the blunt ends of the DNA fragment, to produce a new molecule with sticky ends. This may appear to be a simple method but in practice a new problem arises. The sticky ends of individual adaptor molecules could base pair with each other to form dimers (Figure 4.23(b)), so that the new DNA molecule is still blunt-ended (Figure 4.23(c)). The sticky ends could be recreated by digestion with a restriction endonuclease, but that would defeat the purpose of using adaptors in the first place.

The answer to the problem lies in the precise chemical structure of the ends of the adaptor molecule. Normally the two ends of a polynucleotide

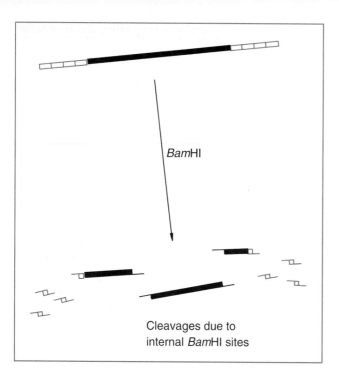

Figure 4.22 A possible problem with the use of linkers. Compare this situation with the desired result of *Bam*HI restriction, as shown in Figure 4.21(b).

BamHI

Cleavages due to internal *Bam*HI sites

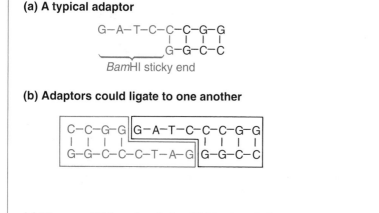

(a) A typical adaptor

G−A−T−C−C−C−G−G
            | | | |
            G−G−C−C

*Bam*HI sticky end

(b) Adaptors could ligate to one another

C−C−G−G G−A−T−C−C−C−G−G
| | | | | | | |
G−G−C−C−C−T−A−G G−G−C−C

Figure 4.23 Adaptors and the potential problem with their use. (a) A typical adaptor. (b) Two adaptors could ligate to one another to produce a molecule similar to a linker, so that (c) after ligation of adaptors a blunt-ended molecule is still blunt-ended and the restriction step is still needed.

(c) The new DNA molecule is still blunt-ended

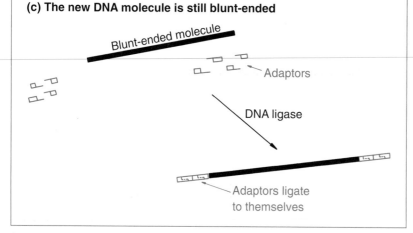

Blunt-ended molecule

Adaptors

DNA ligase

Adaptors ligate to themselves

strand are chemically distinct, a fact that is clear from a careful examination of the polymeric structure of DNA (Figure 4.24(a)). One end, referred to as the 5′ terminus, carries a phosphate group (5′-P); the other, the 3′ terminus, has a hydroxyl group (3′-OH). In the double helix the two strands are antiparallel (Figure 4.24(b)), so each end of a double-stranded molecule consists of one 5′-P terminus and one 3′-OH terminus. Ligation normally takes place between the 5′-P and 3′-OH ends (Figure 4.24(c)).

Figure 4.24 The distinction between the 5′ and 3′ termini of a polynucleotide.

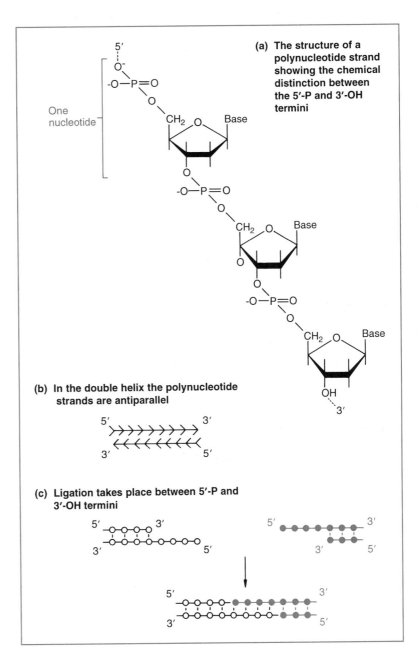

(a) **The structure of a polynucleotide strand showing the chemical distinction between the 5′-P and 3′-OH termini**

(b) **In the double helix the polynucleotide strands are antiparallel**

(c) **Ligation takes place between 5′-P and 3′-OH termini**

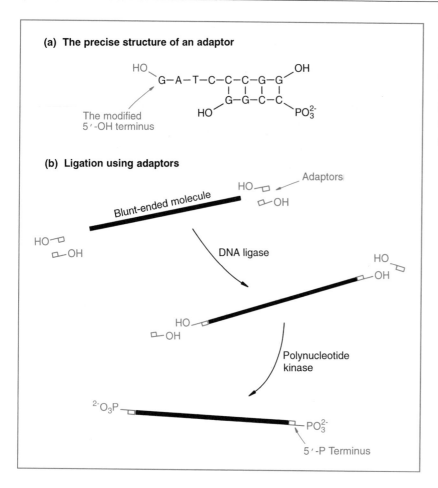

(a) **The precise structure of an adaptor**

HO
G−A−T−C−C−C−G−G   OH
         | | | |
      G−G−C−C
HO   PO₃²⁻

The modified
5′-OH terminus

(b) **Ligation using adaptors**

Adaptors
HO
      OH
Blunt-ended molecule
HO
   OH

DNA ligase
HO
   OH
HO
   OH

Polynucleotide
kinase
²⁻O₃P
                PO₃²⁻
            5′-P Terminus

Figure 4.25 The use of adaptors: (a) The actual structure of an adaptor, showing the modified 5′-OH terminus. (b) Conversion of blunt ends to sticky ends through the attachment of adaptors.

Adaptor molecules are synthesized so that the blunt end is the same as 'natural' DNA, but the sticky end is different. The 3′-OH terminus of the sticky end is the same as usual, but the 5′-P terminus is modified; it lacks the phosphate group, and is in fact a 5′-OH terminus (Figure 4.25(a)). DNA ligase is unable to form a phosphodiester bridge between 5′-OH and 3′-OH ends. The result is that, although base pairing is always occurring between the sticky ends of adaptor molecules, the association is never stabilized by ligation (Figure 4.25(b)).

Adaptors can therefore be ligated to a DNA molecule but not to themselves. After the adaptors have been attached, the abnormal 5′-OH terminus is converted to the natural 5′-P form by treatment with the enzyme polynucleotide kinase (p. 59), producing a sticky-ended fragment that can be inserted into an appropriate vector.

Producing sticky ends by homopolymer tailing

The technique of **homopolymer tailing** offers a radically different approach to the production of sticky ends on a blunt-ended DNA

Figure 4.26 Homopolymer tailing: (a) synthesis of a homopolymer tail, (b) construction of a recombinant DNA molecule from a tailed vector plus tailed insert DNA, and (c) repair of the recombinant DNA molecule. dCTP = 2'-deoxycytidine 5'-triphosphate.

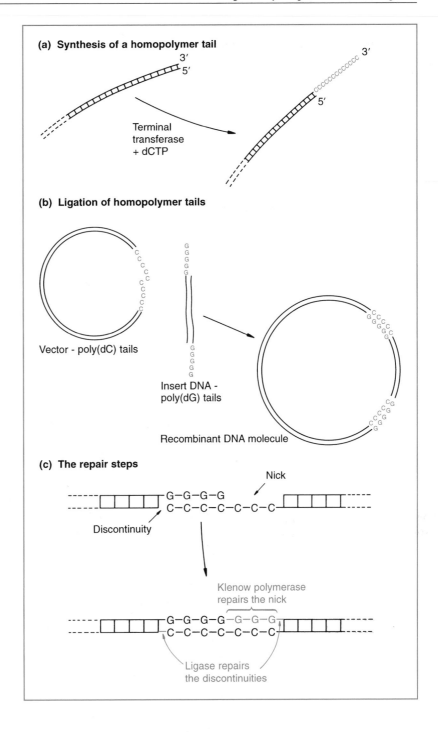

(a) **Synthesis of a homopolymer tail**

Terminal transferase + dCTP

(b) **Ligation of homopolymer tails**

Vector - poly(dC) tails

Insert DNA - poly(dG) tails

Recombinant DNA molecule

(c) **The repair steps**

Nick

G–G–G–G
C–C–C–C–C–C–C

Discontinuity

Klenow polymerase repairs the nick

G–G–G–G–G–G–G–G
C–C–C–C–C–C–C–C

Ligase repairs the discontinuities

molecule. A homopolymer is simply a polymer in which all the subunits are the same. A DNA strand made up entirely of, say, deoxyguanosine is an example of a homopolymer, and is referred to as polydeoxyguanosine or poly(dG).

Tailing involves using the enzyme terminal deoxynucleotidyl transferase (p. 59) to add a series of nucleotides onto the 3'-OH termini of a double-stranded DNA molecule. If this reaction is carried out in the presence of just one deoxyribonucleotide a homopolymer tail is produced (Figure 4.26(a)).

Of course, to be able to ligate together two tailed molecules, the homopolymers must be complementary, Frequently polydeoxycytosine (poly(dC)) tails are attached to the vector and poly(dG) to the DNA to be cloned. Base pairing between the two occurs when the DNA molecules are mixed (Figure 4.26(b)).

In practice the poly(dG) and poly(dC) tails are not usually exactly the same length, and the base-paired recombinant molecules that result have nicks as well as discontinuities (Figure 4.26(c)). Repair is therefore a two step process, using Klenow polymerase to fill in the nicks followed by DNA ligase to synthesize the final phosphodiester bonds. This repair reaction does not always have to be performed in the test tube. If the complementary homopolymer tails are longer than about 20 nucleotides, then quite stable base-paired associations are formed. A recombinant DNA molecule, held together by base pairing although not completely ligated, is often stable enough to be introduced into the host cell in the next stage of the cloning experiment (Figure 1.1). Once inside the host, the cell's own DNA polymerase and DNA ligase repair the recombinant DNA molecule, completing the construction begun in the test tube.

Further reading

Brown, T.A. (1998) *Molecular Biology Labfax. Volume I: Recombinant DNA*, 2nd edn. Academic Press, London. [Contains details of all types of enzymes used to manipulate DNA and RNA.]

Jacobsen, H., Klenow, H. & Overgaard-Hansen, K. (1974) The N-terminal amino acid sequences of DNA polymerase I from *Escherichia coli* and of the large and small fragments obtained by a limited proteolysis. *European Journal of Biochemistry*, **45**, 623–7. [Production of the Klenow fragment of DNA polymerase I.]

Lobban, P. & Kaiser, A.D. (1973) Enzymatic end-to-end joining of DNA molecules. *Journal of Molecular Biology*, **79**, 453–71. [Ligation.]

McDonell, M.W., Simon, M.N. & Studier, F.W. (1977) Analysis of restriction fragments of T7 DNA and determination of molecular weights by electrophoresis in neutral and alkaline gels. *Journal of Molecular Biology*, **110**, 119–46. [An early example of the use of agarose gel electrophoresis in the analysis of restriction fragment sizes.]

REBASE: www.neb.com/rebase/rebase.html [A comprehensive list of all the known restriction endonucleases and their recognition sequences.]

Rothstein, R.J., Lau, L.F., Bahl, C.P., Narang, N.A. & Wu, R. (1979) Synthetic adaptors for cloning DNA. *Methods in Enzymology*, **68**, 98–109.

Smith, H.O. & Wilcox, K.W. (1970) A restriction enzyme from *Haemophilus influenzae*. *Journal of Molecular Biology*, **51**, 379–91. [One of the first full descriptions of a restriction endonuclease.]

Chapter 5

Introduction of DNA into Living Cells

The manipulations described in Chapter 4 allow the molecular biologist to create novel recombinant DNA molecules. The next step in a gene cloning experiment is to introduce these molecules into living cells, usually bacteria, which then grow and divide to produce clones (Figure 1.1). Strictly speaking, the word 'cloning' refers only to the later stages of the procedure, and not to the construction of the recombinant DNA molecule itself.

Cloning serves two main purposes. First, it allows a large number of recombinant DNA molecules to be produced from a limited amount of starting material. At the outset only a few nanograms of recombinant DNA may be available, but each bacterium that takes up a plasmid subsequently divides numerous times to produce a colony, each cell of which contains multiple copies of the molecule. Several micrograms of recombinant DNA can usually be prepared from a single bacterial colony, representing a thousandfold increase over the starting amount (Figure 5.1). If the colony is used not as a source of DNA but as an inoculum for a liquid culture, the resulting cells may provide milligrams of DNA, a millionfold increase in yield. In this way cloning can supply the large amounts of DNA needed for molecular biological studies of gene structure and expression (Chapters 10 and 11).

The second important function of cloning can be described as purification. The manipulations that result in a recombinant DNA molecule can only rarely be controlled to the extent that no other DNA molecules are present at the end of the procedure. The ligation mixture may contain, in addition to the desired recombinant molecule, any number of the following (Figure 5.2(a)):

(1) Unligated vector molecules.
(2) Unligated DNA fragments.
(3) Vector molecules that have recircularized without new DNA being inserted ('self-ligated' vector).

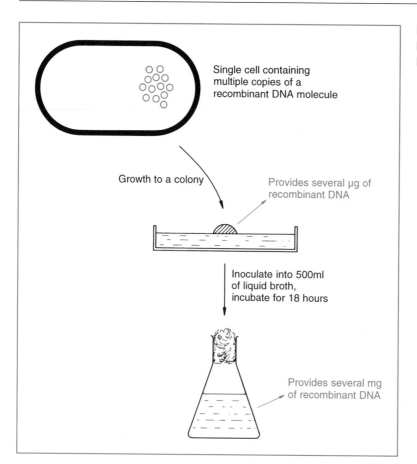

Single cell containing
multiple copies of a
recombinant DNA molecule

Growth to a colony

Provides several µg of
recombinant DNA

Inoculate into 500ml
of liquid broth,
incubate for 18 hours

Provides several mg
of recombinant DNA

Figure 5.1 Cloning can supply large amounts of recombinant DNA.

(4) Recombinant DNA molecules that carry the wrong inserted DNA fragment.

Unligated molecules rarely cause a problem because, even though they may be taken up by bacterial cells, only under exceptional circumstances will they be replicated. It is much more likely that enzymes within the host bacteria degrade these pieces of DNA. On the other hand, self-ligated vector molecules and incorrect recombinant plasmids are replicated just as efficiently as the desired molecule (Figure 5.2(b)). However, purification of the desired molecule can still be achieved through cloning because it is extremely unusual for any one cell to take up more than one DNA molecule. Each cell gives rise to a single colony, so each of the resulting clones consists of cells that all contain the same molecule. Of course different colonies contain different molecules: some contain the desired recombinant DNA molecule, some have different recombinant molecules, and some contain self-ligated vector. The problem therefore becomes a question of identifying the colonies that contain the correct recombinant plasmids.

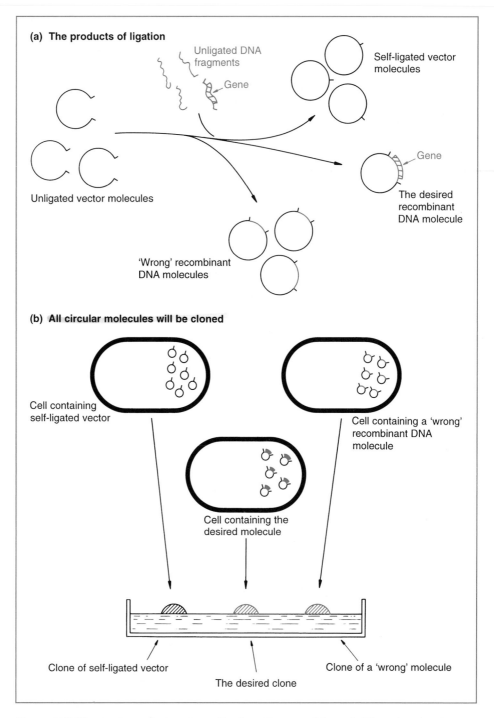

Figure 5.2 Cloning is analogous to purification. From a mixture of different molecules, clones containing copies of just one molecule can be obtained.

This chapter is concerned with the way in which plasmid and phage vectors, and recombinant molecules derived from them, are introduced into bacterial cells. During the course of the chapter it will become apparent that selection for colonies containing recombinant molecules, as opposed to colonies containing self-ligated vector, is relatively easy. The more difficult proposition of how to distinguish clones containing the correct recombinant DNA molecule from all the other recombinant clones will be tackled in Chapter 8.

5.1 Transformation – the uptake of DNA by bacterial cells

Most species of bacteria are able to take up DNA molecules from the medium in which they grow. Often a DNA molecule taken up in this way will be degraded, but occasionally it is able to survive and replicate in the host cell. In particular this happens if the DNA molecule is a plasmid with an origin of replication recognized by the host.

Uptake and stable retention of a plasmid is usually detected by looking for expression of the genes carried by the plasmid (p. 14). For example, *E. coli* cells are normally sensitive to the growth inhibitory effects of the antibiotics ampicillin and tetracycline. However, cells that contain the plasmid pBR322 (p. 105), one of the first cloning vectors to be developed back in the 1970s, are resistant to these antibiotics. This is because pBR322 carries two sets of genes, one gene that codes for a β-lactamase enzyme that modifies ampicillin into a form that is non-toxic to the bacterium, and a second set of genes that code for enzymes that detoxify tetracycline. Uptake of pBR322 can be detected because the *E. coli* cells are transformed from ampicillin and tetracycline sensitive ($amp^S tet^S$) to ampicillin and tetracycline resistant ($amp^R tet^R$).

In recent years the term **transformation** has been extended to include uptake of any DNA molecule by any type of cell, regardless of whether the uptake results in a detectable change in the cell, or whether the cell involved is bacterial, fungal, animal or plant.

5.1.1 Not all species of bacteria are equally efficient at DNA uptake

In nature, transformation is probably not a major process by which bacteria obtain genetic information. This is reflected by the fact that in the laboratory only a few species (notably members of the genera *Bacillus* and *Streptococcus*) can be transformed with ease. Close study of these organisms has revealed that they possess sophisticated mechanisms for DNA binding and uptake.

Most species of bacteria, including *E. coli*, take up only limited amounts of DNA under normal circumstances. In order to transform these species efficiently, the bacteria have to undergo some form of physical and/or chemical treatment that enhances their ability to take up DNA. Cells that have undergone this treatment are said to be **competent**.

5.1.2 Preparation of competent *E. coli* cells

As with many breakthroughs in recombinant DNA technology, the key development as far as transformation is concerned occurred in the early 1970s, when it was observed that *E. coli* cells that had been soaked in an ice cold salt solution were more efficient at DNA uptake than unsoaked cells. A solution of 50 mM calcium chloride ($CaCl_2$) is traditionally used, although other salts, notably rubidium chloride, are also effective.

Exactly why this treatment works is not understood. Possibly $CaCl_2$ causes the DNA to precipitate onto the outside of the cells, or perhaps the salt is responsible for some kind of change in the cell wall that improves DNA binding. In any case, soaking in $CaCl_2$ affects only DNA binding, and not the actual uptake into the cell. When DNA is added to treated cells, it remains attached to the cell exterior, and is not at this stage transported into the cytoplasm (Figure 5.3). The actual movement of DNA into competent cells is stimulated by briefly raising the temperature

Figure 5.3 The binding and uptake of DNA by a competent bacterial cell.

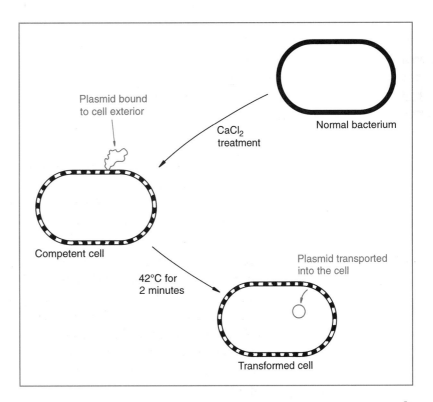

Plasmid bound to cell exterior

$CaCl_2$ treatment

Normal bacterium

Competent cell

Plasmid transported into the cell

42°C for 2 minutes

Transformed cell

to 42°C. Once again, the exact reason why this heat shock is effective is not understood.

Selection for transformed cells

Transformation of competent cells is an inefficient procedure, however carefully the cells have been prepared. Although 1 ng of the plasmid vector called pUC8 (p. 108) can yield 1000–10000 transformants, this represents the uptake of only 0.01% of all the available molecules. Furthermore, 10000 transformants is only a very small proportion of the total number of cells that are present in a competent culture. This last fact means that some way must be found to distinguish a cell that has taken up a plasmid from the many thousands that have not been transformed.

The answer is to make use of a selectable marker carried by the plasmid. A selectable marker is simply a gene that provides a transformed cell with a new characteristic, one that is not possessed by a non-transformant. A good example of a selectable marker is the ampicillin resistance gene of pBR322. After a transformation experiment with pBR322, only those *E. coli* cells that have taken up a plasmid are $amp^R tet^R$ and able to form colonies on an agar medium that contains ampicillin or tetracycline (Figure 5.4); non-transformants, which are still $amp^S tet^S$, do not produce colonies on the

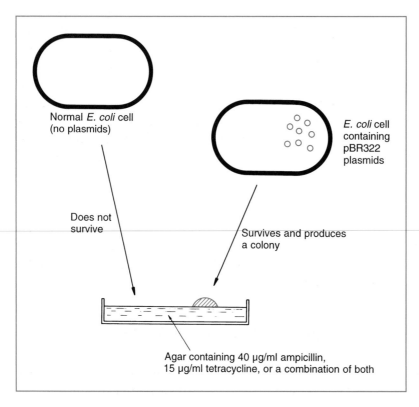

Figure 5.4 Selecting cells that contain pBR322 plasmids by plating onto agar medium containing ampicillin and/or tetracycline.

Normal *E. coli* cell (no plasmids)

E. coli cell containing pBR322 plasmids

Does not survive

Survives and produces a colony

Agar containing 40 µg/ml ampicillin, 15 µg/ml tetracycline, or a combination of both

Figure 5.5 Phenotypic expression. Incubation at 37°C for 1 hour before plating out improves the survival of the transformants on selective medium, because the bacteria have had time to begin synthesis of the antibiotic resistance enzymes.

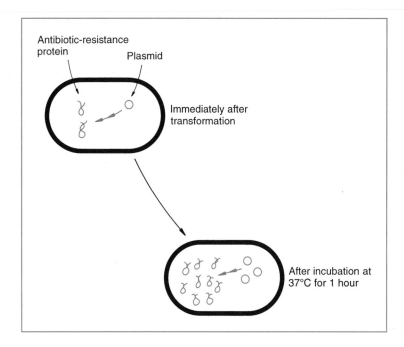

selective medium. Transformants and non-transformants are therefore easily distinguished.

Most plasmid cloning vectors carry at least one gene that confers antibiotic resistance on the host cells, with selection of transformants being achieved by plating onto an agar medium that contains the relevant antibiotic. Bear in mind, however, that resistance to the antibiotic is not due merely to the presence of the plasmid in the transformed cells. The resistance gene on the plasmid must also be expressed, so that the enzyme that detoxifies the antibiotic is synthesized. Expression of the resistance gene begins immediately after transformation, but it will be a few minutes before the cell contains enough of the enzyme to be able to withstand the toxic effects of the antibiotic. For this reason the transformed bacteria should not be plated onto the selective medium immediately after the heat shock treatment, but first placed in a small volume of liquid medium, in the absence of antibiotic, and incubated for a short time. Plasmid replication and expression can then get started, so that when the cells are plated out and encounter the antibiotic, they will already have synthesized sufficient resistance enzymes to be able to survive (Figure 5.5).

5.2 Identification of recombinants

Plating onto a selective medium enables transformants to be distinguished from non-transformants. The next problem is to determine which of the

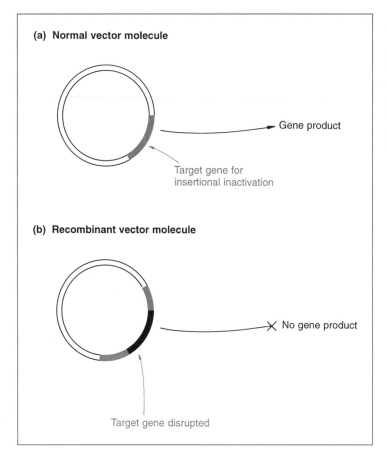

(a) Normal vector molecule

Gene product

Target gene for
insertional inactivation

(b) Recombinant vector molecule

No gene product

Target gene disrupted

Figure 5.6 Insertional inactivation. (a) The normal, non-recombinant vector molecule carries a gene whose product confers a selectable or identifiable characteristic on the host cell. (b) This gene is disrupted when new DNA is inserted into the vector; as a result the recombinant host does not display the relevant characteristic.

transformed colonies comprise cells that contain recombinant DNA molecules, and which contain self-ligated vector molecules (Figure 5.2). With most cloning vectors insertion of a DNA fragment into the plasmid destroys the integrity of one of the genes present on the molecule. **Recombinants** can therefore be identified because the characteristic coded by the inactivated gene is no longer displayed by the host cells (Figure 5.6). The general principles of **insertional inactivation** are illustrated by a typical cloning experiment using pBR322 as the vector.

5.2.1

Recombinant selection with pBR322 – insertional inactivation of an antibiotic resistance gene

pBR322 has several unique restriction sites that can be used to open up the vector before insertion of a new DNA fragment (Figure 5.7(a)). *Bam*HI, for example, cuts pBR322 at just one position, within the cluster of genes that code for resistance to tetracycline. A recombinant pBR322 molecule, one that carries an extra piece of DNA in the *Bam*HI site (Figure 5.7(b)), is no longer

Figure 5.7 The cloning vector pBR322: (a) the normal vector molecule, and (b) a recombinant molecule containing an extra piece of DNA inserted into the *Bam*HI site. For a more detailed map of pBR322 see Figure 6.1.

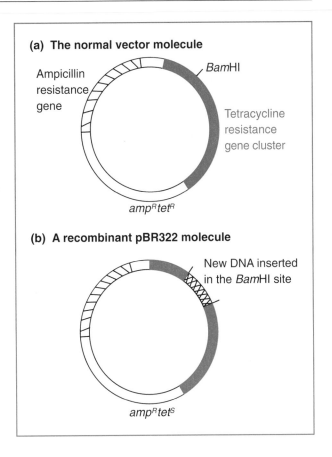

(a) The normal vector molecule

Ampicillin resistance gene

*Bam*HI

Tetracycline resistance gene cluster

amp^R*tet*^{R}

(b) A recombinant pBR322 molecule

New DNA inserted in the *Bam*HI site

amp^R*tet*^{S}

able to confer tetracycline resistance on its host, as one of the necessary genes is now disrupted by the inserted DNA. Cells containing this recombinant pBR322 molecule are still resistant to ampicillin, but sensitive to tetracycline ($amp^R tet^S$).

Screening for pBR322 recombinants is performed in the following way. After transformation the cells are plated onto ampicillin medium and incubated until colonies appear (Figure 5.8(a)). All of these colonies are transformants (remember, untransformed cells are amp^S and so do not produce colonies on the selective medium), but only a few contain recombinant pBR322 molecules: most contain the normal, self-ligated plasmid. To identify the recombinants the colonies are **replica plated** onto agar medium that contains tetracycline (Figure 5.8(b)). After incubation, some of the original colonies regrow, but others do not (Figure 5.8(c)). Those that do grow consist of cells that carry the normal pBR322 with no inserted DNA and therefore a functional tetracycline resistance gene cluster ($amp^R tet^R$). The colonies that do not grow on tetracycline agar are recombinants ($amp^R tet^S$); once their positions are known, samples for further study can be recovered from the original ampicillin agar plate.

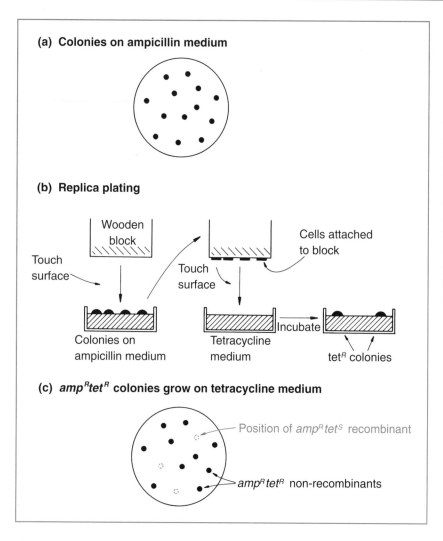

(a) Colonies on ampicillin medium

(b) Replica plating

(c) $amp^R tet^R$ **colonies grow on tetracycline medium**

Figure 5.8 Screening for pBR322 recombinants by insertional inactivation of the tetracycline resistance gene. (a) Cells are plated onto ampicillin agar: all the transformants produce colonies. (b) The colonies are replica plated onto tetracycline medium. (c) The colonies that grow on tetracycline medium are $amp^R tet^R$ and therefore non-recombinants. Recombinants ($amp^R tet^S$) do not grow, but their position on the ampicillin plate is now known.

5.2.2 Insertional inactivation does not always involve antibiotic resistance

Although insertional inactivation of an antibiotic resistance gene provides a convenient means of recombinant identification, several important plasmid cloning vectors make use of a different system. An example is pUC8 (Figure 5.9(a)), which carries the ampicillin resistance gene and a gene called *lacZ'*, which codes for part of the enzyme β-galactosidase. Cloning with pUC8 involves insertional inactivation of the *lacZ'* gene, with recombinants identified because of their inability to synthesize β-galactosidase (Figure 5.9(b)).

β-galactosidase is one of a series of enzymes involved in the breakdown of lactose to glucose plus galactose. It is normally coded by the gene *lacZ*, which resides on the *E. coli* chromosome. Some strains of *E. coli* have a

Figure 5.9 The cloning vector pUC8: (a) the normal vector molecule, (b) a recombinant molecule containing an extra piece of DNA inserted into the *Bam*HI site. For more detailed maps of pUC8 see Figures 6.3 and 6.4.

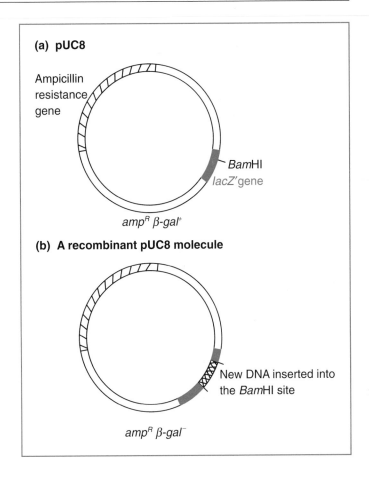

modified *lacZ* gene, one that lacks the segment referred to as *lacZ'* and coding for the α-peptide portion of β-galactosidase (Figure 5.10(a)). These mutants can synthesize the enzyme only when they harbour a plasmid, such as pUC8, that carries the missing *lacZ'* segment of the gene.

A cloning experiment with pUC8 involves selection of transformants on ampicillin agar followed by screening for β-galactosidase activity to identify recombinants. Cells that harbour a normal pUC8 plasmid are *amp*^{R} and able to synthesize β-galactosidase (Figure 5.9(a)); recombinants are also *amp*^{R} but unable to make β-galactosidase (Figure 5.9(b)).

Screening for β-galactosidase presence or absence is in fact quite easy. Rather than assay for lactose being split to glucose and galactose, a slightly different reaction catalysed by the enzyme is tested for. This involves a lactose analogue called X-gal (5-bromo-4-chloro-3-indolyl-β-D-galactopyranoside) which is broken down by β-galactosidase to a product that is coloured deep blue. If X-gal (plus an inducer of the enzyme such as isopropylthiogalactoside, IPTG) is added to the agar, along with ampicillin, then non-recombinant colonies, the cells of which synthesize β-galactosidase, will

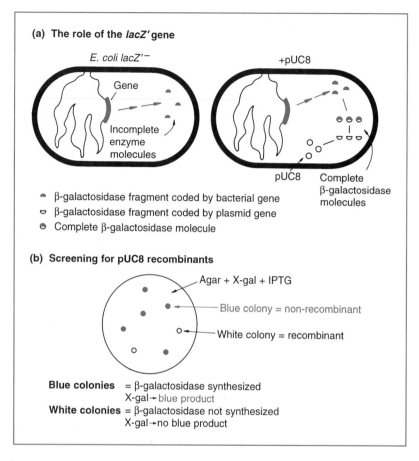

(a) The role of the *lacZ'* gene

E. coli lacZ'⁻

+pUC8

Gene

Incomplete enzyme molecules

pUC8   Complete β-galactosidase molecules

- β-galactosidase fragment coded by bacterial gene
- β-galactosidase fragment coded by plasmid gene
- Complete β-galactosidase molecule

(b) Screening for pUC8 recombinants

Agar + X-gal + IPTG

Blue colony = non-recombinant

White colony = recombinant

Blue colonies = β-galactosidase synthesized
X-gal → blue product
White colonies = β-galactosidase not synthesized
X-gal → no blue product

Figure 5.10 The rationale behind insertional inactivation of the *lacZ'* gene carried by pUC8. (a) The bacterial and plasmid genes complement each other to produce a functional β-galactosidase molecule. (b) Recombinants are screened by plating onto agar containing X-gal and IPTG.

be coloured blue, whereas recombinants with a disrupted *lacZ'* gene and unable to make β-galactosidase, will be white. This system, which is called **Lac selection**, is summarized in Figure 5.10(b).

5.3 Introduction of phage DNA into bacterial cells

There are two different methods by which a recombinant DNA molecule constructed with a phage vector can be introduced into a bacterial cell: transfection and *in vitro* packaging.

5.3.1 Transfection

This process is equivalent to transformation, the only difference being that phage DNA rather than a plasmid is involved. Just as with a plasmid, the purified phage DNA, or recombinant phage molecule, is mixed with competent *E. coli* cells and DNA uptake induced by heat shock.

5.3.2 *In vitro* packaging

Transfection with λ DNA molecules is not a very efficient process when compared with the infection of a culture of cells with mature λ phage particles. It would therefore be useful if recombinant λ molecules could be packaged into their λ head-and-tail structures in the test tube.

This may sound difficult but is actually relatively easy to achieve. Packaging requires a number of different proteins coded by the λ genome, but these can be prepared at a high concentration from cells infected with defective λ phage strains. Two different systems are in use. With the single strain system the defective λ phage carries a mutation in the *cos* sites, so that these are not recognized by the endonuclease that normally cleaves the λ catenanes during phage replication (p. 23). The defective phage cannot therefore replicate, though it does direct synthesis of all the proteins needed for packaging. The proteins accumulate in the bacterium and can be purified from cultures of *E. coli* infected with the mutated λ, and used for *in vitro* packaging of recombinant λ molecules (Figure 5.11(a)).

With the second system two defective λ strains are needed. Both of these strains carry a mutation in a gene for one of the components of the phage protein coat: with one strain the mutation is in gene *D*, and with the second strain it is in gene *E* (Figure 2.9). Neither strain is able to complete an infection cycle in *E. coli* because in the absence of the product of the mutated gene the complete capsid structure cannot be made. Instead the products of all the other coat protein genes accumulate (Figure 5.11(b)). An *in vitro* packaging mix can therefore be prepared by combining lysates of two cultures of cells, one infected with the λ D^- strain, the other infected with the E^- strain. The mixture now contains all the necessary components and can package recombinant λ molecules into mature phage particles (Figure 5.11(c)).

With both systems, packaged molecules are introduced into *E. coli* cells simply by adding the assembled phage to the bacterial culture and allowing the normal λ infective process to take place.

5.3.3 Phage infection is visualized as plaques on an agar medium

The final stage of the phage infection cycle is cell lysis (p. 20). If infected cells are spread onto a solid agar medium immediately after addition of the phage, or immediately after transfection with phage DNA, cell lysis can be visualized as **plaques** on a lawn of bacteria (Figure 5.12(a)). Each plaque is a zone of clearing produced as the phages lyse the cells and move on to infect and eventually lyse the neighbouring bacteria (Figure 5.12(b)).

Both λ and M13 form plaques. With λ these are true plaques, produced by cell lysis. However, M13 plaques are slightly different as M13 does not lyse the host cells (p. 21). Instead M13 causes a decrease in the growth

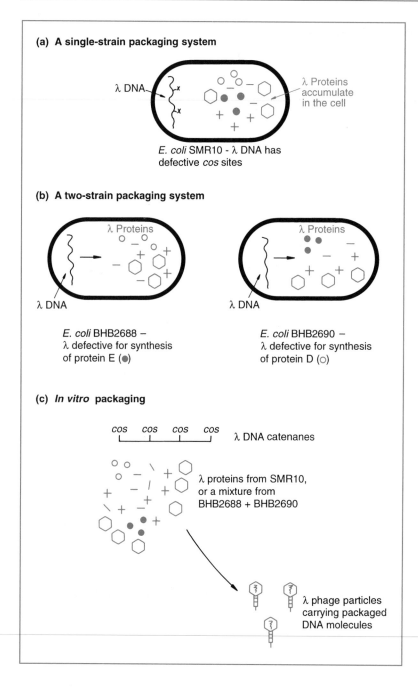

(a) A single-strain packaging system

λ DNA

λ Proteins accumulate in the cell

E. coli SMR10 - λ DNA has defective *cos* sites

(b) A two-strain packaging system

λ Proteins

λ DNA

E. coli BHB2688 – λ defective for synthesis of protein E (●)

λ Proteins

λ DNA

E. coli BHB2690 – λ defective for synthesis of protein D (○)

(c) *In vitro* packaging

cos *cos* *cos* *cos*

λ DNA catenanes

λ proteins from SMR10, or a mixture from BHB2688 + BHB2690

λ phage particles carrying packaged DNA molecules

Figure 5.11 *In vitro* packaging. (a) Synthesis of λ capsid proteins by *E. coli* strain SMR10, which carries a λ phage that has defective *cos* sites. (b) Synthesis of incomplete sets of λ capsid proteins by *E. coli* strains BHB2688 and BHB2690. (c) A mixture of cell lysates provides the complete set of capsid proteins and can package λ DNA molecules in the test tube.

rate of infected cells, sufficient to produce a zone of relative clearing on a bacterial lawn. Although not true plaques, these zones of clearing are visually identical to normal phage plaques (Figure 5.12(c)).

The end result of a gene cloning experiment using a λ or M13 vector is therefore an agar plate covered in phage plaques. Each plaque is derived from a single transfected or infected cell and therefore contains identical

Figure 5.12 Bacteriophage plaques. (a) The appearance of plaques on a lawn of bacteria. (b) Plaques produced by a phage that lyses the host cell (e.g. λ in the lytic infection cycle); the plaques contain lysed cells plus many phage particles. (c) Plaques produced by M13; these plaques contain slow-growing bacteria plus many M13 phage particles.

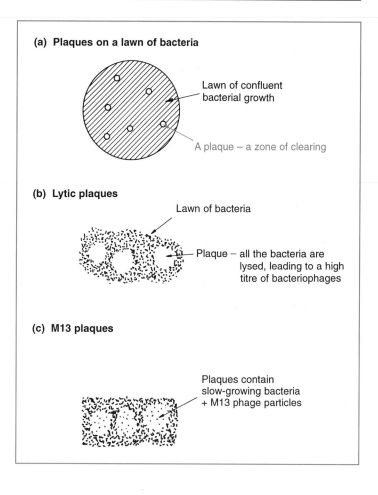

(a) Plaques on a lawn of bacteria

Lawn of confluent bacterial growth

A plaque – a zone of clearing

(b) Lytic plaques

Lawn of bacteria

Plaque – all the bacteria are lysed, leading to a high titre of bacteriophages

(c) M13 plaques

Plaques contain slow-growing bacteria + M13 phage particles

phage particles. These may contain self-ligated vector molecules, or they may be recombinants.

5.4   Identification of recombinant phages

A variety of ways of distinguishing recombinant plaques have been devised, the following being the most important.

5.4.1   Insertional inactivation of a *lacZ'* gene carried by the phage vector

All M13 cloning vectors (p. 111), as well as several λ vectors, carry a copy of the *lacZ'* gene. Insertion of new DNA into this gene inactivates β-galactosidase synthesis, just as with the plasmid vector pUC8. Recombinants are distinguished by plating cells onto X-gal agar: plaques comprising normal phage are blue, recombinant plaques are clear (Figure 5.13(a)).

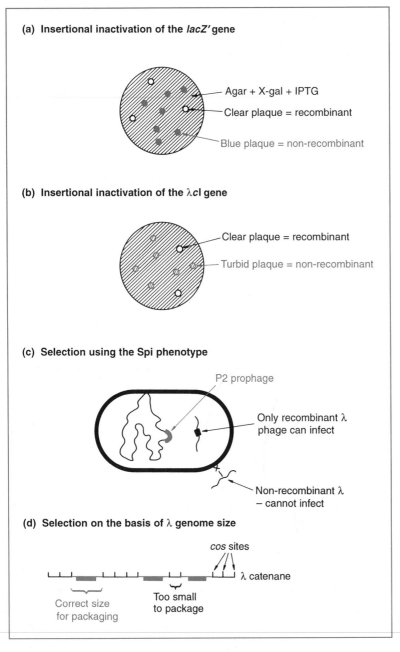

(a) **Insertional inactivation of the *lacZ'* gene**

— Agar + X-gal + IPTG

— Clear plaque = recombinant

— Blue plaque = non-recombinant

(b) **Insertional inactivation of the λ*cI* gene**

— Clear plaque = recombinant

— Turbid plaque = non-recombinant

(c) **Selection using the Spi phenotype**

P2 prophage

Only recombinant λ phage can infect

Non-recombinant λ – cannot infect

(d) **Selection on the basis of λ genome size**

cos sites

λ catenane

Correct size for packaging

Too small to package

Figure 5.13 Strategies for the selection of recombinant phage.

5.4.2

Insertional inactivation of the λ *cI* gene

Several types of λ cloning vector have unique restriction sites in the *cI* gene (map position 38 on Figure 2.9). Insertional inactivation of this gene causes a change in plaque morphology. Normal plaques appear 'turbid', whereas recombinants with a disrupted *cI* gene are 'clear' (Figure 5.13(b)). The difference is readily apparent to the experienced eye.

5.4.3

Selection using the Spi phenotype

λ phage cannot normally infect *E. coli* cells that already possess an integrated form of a related phage called P2. λ is therefore said to be Spi⁻ (sensitive to P2 prophage inhibition). Some λ cloning vectors are designed so that insertion of new DNA causes a change from Spi⁺ to Spi⁻, enabling the recombinants to infect cells that carry P2 prophages. Such cells are used as the host for cloning experiments with these vectors; only recombinants are Spi⁻ so only recombinants form plaques (Figure 5.13(c)).

5.4.4

Selection on the basis of λ genome size

The λ packaging system, which assembles the mature phage particles, can only insert DNA molecules of between 37 and 52 kb into the head structure. Anything less than 37 kb is not packaged. Many λ vectors have been constructed by deleting large segments of the λ DNA molecule (p. 120) and so are less than 37 kb in length. These can only be packaged into mature phage particles after extra DNA has been inserted, bringing the total genome size up to 37 kb or more (Figure 5.13(d)). Therefore with these vectors only recombinant phage are able to replicate.

5.5

Transformation of non-bacterial cells

Ways of introducing DNA into yeast, fungi, animals and plants are also needed if these organisms are to be used as the hosts for gene cloning. In general terms, soaking cells in salt is effective only with a few species of bacteria, although treatment with lithium chloride or lithium acetate does enhance DNA uptake by yeast cells, and is frequently used in the transformation of *Saccharomyces cerevisiae*. However, for most higher organisms more sophisticated methods are needed.

5.5.1

Transformation of individual cells

With most organisms the main barrier to DNA uptake is the cell wall. Cultured animal cells, which usually lack cell walls, are easily transformed, especially if the DNA is precipitated onto the cell surface with calcium phosphate (Figure 5.14(a)). For other types of cell the answer is often to remove the cell wall. Enzymes that degrade yeast, fungal and plant cell walls are available, and under the right conditions intact **protoplasts** can be obtained (Figure 5.14(b)). Protoplasts generally take up DNA quite readily, but transformation can be stimulated by special techniques such as **electroporation**, during which the cells are subjected to a short electrical pulse, thought to induce the transient formation of pores in the cell membrane, through which DNA molecules are able to enter the cell. After transformation the protoplasts are washed to remove the degradative enzymes and the cell wall spontaneously re-forms.

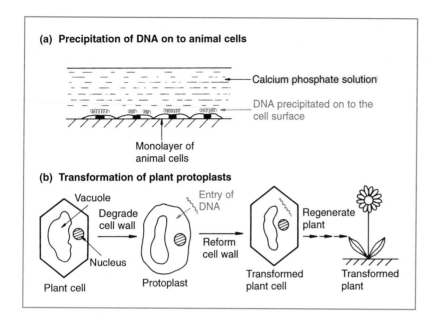

Figure 5.14 Strategies for introducing new DNA into animal and plant cells.

(a) Precipitation of DNA on to animal cells

Calcium phosphate solution

DNA precipitated on to the cell surface

Monolayer of animal cells

(b) Transformation of plant protoplasts

Vacuole

Degrade cell wall

Nucleus

Plant cell

Protoplast

Entry of DNA

Reform cell wall

Transformed plant cell

Regenerate plant

Transformed plant

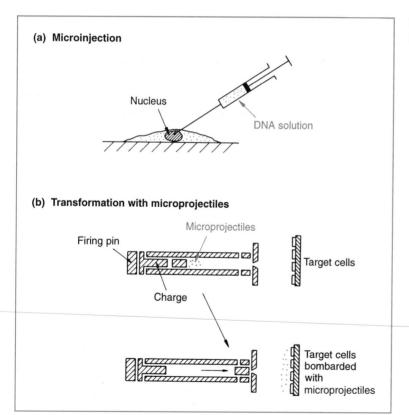

Figure 5.15 Two physical methods for introducing DNA into cells.

(a) Microinjection

Nucleus

DNA solution

(b) Transformation with microprojectiles

Microprojectiles

Firing pin

Charge

Target cells

Target cells bombarded with microprojectiles

In contrast to the transformation systems described so far, there are two physical methods for introducing DNA into cells. The first of these is **microinjection**, which makes use of a very fine pipette to inject DNA molecules directly into the nucleus of the cells to be transformed (Figure 5.15(a)). This technique was initially applied to animal cells but has subsequently been successful with plant cells. The second method involves bombardment of the cells with high velocity microprojectiles, usually particles of gold or tungsten that have been coated with DNA. These microprojectiles are fired at the cells from a particle gun (Figure 5.15(b)). This unusual technique is termed **biolistics** and has been used with a number of different types of cell.

5.5.2   Transformation of whole organisms

With animals and plants the desired end product may not be transformed cells, but a transformed organism. Plants are relatively easy to regenerate from cultured cells, though problems have been experienced in developing regeneration procedures for monocotyledonous species such as cereals and grasses. A single transformed plant cell can therefore give rise to a transformed plant, which carries the cloned DNA in every cell, and passes the cloned DNA on to its progeny following flowering and seed formation (see Figure 7.13(b)). Animals, of course, cannot be regenerated from cultured cells, so obtaining transformed animals requires a rather more subtle approach. One technique with mammals such as mice is to remove fertilized eggs from the oviduct, to microinject DNA, and then to reimplant the transformed cells into the mother's reproductive tract (p. 153).

Further reading

Calvin, N.M. & Hanawalt, P.C. (1988) High efficiency transformation of bacterial cells by electroporation. *Journal of Bacteriology*, **170**, 2796–801.

Capecchi, M.R. (1980) High efficiency transformation by direct microinjection of DNA into cultured mammalian cells. *Cell*, **22**, 479–88.

Cohen, S.N., Chang, A.V.Y., Hsu, L. *et al.* (1972) Nonchromosomal antibiotic resistance in bacteria: genetic transformation of *Escherichia coli* by R-factor DNA. *Proceedings of the National Academy of Sciences of the USA*, **69**, 2110–14. [Transformation of a bacterium with a plasmid.]

Hammer, R.E., Pursel, V.G., Rexroad, C.E. *et al.* (1985) Production of transgenic rabbits, sheep and pigs by microinjection. *Nature*, **315**, 680–83.

Hohn, B. & Murray, K. (1977) Packaging recombinant DNA molecules into bacteriophage particles *in vitro*. *Proceedings of the National Academy of Sciences of the USA*, **74**, 3259–63. [*In vitro* packaging.]

Klein, T.M., Wolf, E.D., Wu, R. & Sanford, J.C. (1987) High velocity microprojectiles for delivering nucleic acids into living cells. *Nature*, **327**, 70–73. [Biolistics.]

Mandel, M. & Higa, A. (1970) Calcium-dependent bacteriophage DNA infection. *Journal of Molecular Biology*, **53**, 154–62. [Transfection.]

recombinant DNA molecules constructed with it. Even with 6 kb of additional DNA, a recombinant pBR322 molecule is still a manageable size.

The second feature of pBR322 is that, as described in Chapter 5, it carries two sets of antibiotic resistance genes. Either ampicillin or tetracycline resistance can be used as a selectable marker for cells containing the plasmid, and each marker gene includes unique restriction sites that can be used in cloning experiments. Insertion of new DNA into pBR322 that has been restricted with *Pst*I, *Pvu*I or *Sca*I inactivates the *amp*^R gene, and insertion using any one of eight restriction endonucleases (notably *Bam*HI and *Hin*dIII) inactivates tetracycline resistance. This great variety of restriction sites that can be used for insertional inactivation means that pBR322 can be used to clone DNA fragments with any of several kinds of sticky end.

A third advantage of pBR322 is that it has a reasonably high copy number. Generally there are about 15 molecules present in a transformed *E. coli* cell, but this number can be increased, up to 1000–3000, by plasmid amplification in the presence of a protein synthesis inhibitor such as chloramphenicol (p. 44). An *E. coli* culture therefore provides a good yield of recombinant pBR322 molecules.

6.1.3 The pedigree of pBR322

The remarkable convenience of pBR322 as a cloning vector did not arise by chance. The plasmid was in fact designed in such a way that the final construct would possess these desirable properties. An outline of the scheme used to construct pBR322 is shown in Figure 6.2(a). It can be seen that its production was a tortuous business that required full and skilful use of the DNA manipulative techniques described in Chapter 4. A summary of the result of these manipulations is provided in Figure 6.2(b), from which it can be seen that pBR322 in fact comprises DNA derived from three different naturally occurring plasmids. The *amp*^R gene originally resided on the plasmid R1, a typical antibiotic resistance plasmid that occurs in natural populations of *E. coli* (p. 18). The *tet*^R gene is derived from R6-5, a second antibiotic resistance plasmid. The replication origin of pBR322, which directs multiplication of the vector in host cells, is originally from pMB1, which is closely related to the colicin-producing plasmid ColE1 (p. 18).

6.1.4 Other typical *E. coli* plasmid cloning vectors

pBR322 was developed in the late 1970s, the first research paper describing its use being published in 1977. Since then many other plasmid cloning vectors have been constructed, the majority of these derived from pBR322 by manipulations similar to those summarized in Figure 6.2(a). It would be pointless to attempt to describe all of these vectors, especially as many are variations on a common theme. Three additional examples will suffice to illustrate the most important features displayed by the range of plasmid cloning vectors available to today's genetic engineers.

Figure 6.2 The pedigree of pBR322: (a) the manipulations involved in construction of pBR322, and (b) a summary of the origins of pBR322.

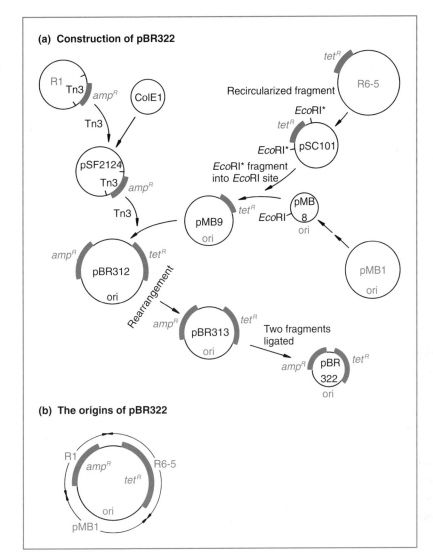

pBR327 – a higher copy number plasmid

pBR327 (Figure 6.3(a)) was constructed by removing a 1089 bp segment from pBR322. This deletion left the *amp*^{R} and *tet*^{R} genes intact, but changed the replicative and conjugative abilities of the resulting plasmid. As a result pBR327 differs from pBR322 in two important ways.

(1) pBR327 has a higher copy number than pBR322, being present at about 30–45 molecules per *E. coli* cell. This is not of great relevance as far as plasmid yield is concerned, as both plasmids can be amplified to copy numbers greater than 1000. However, the higher copy number of pBR327 in normal cells makes this vector more suitable if the aim of the experiment is to study the function of the cloned gene. In these

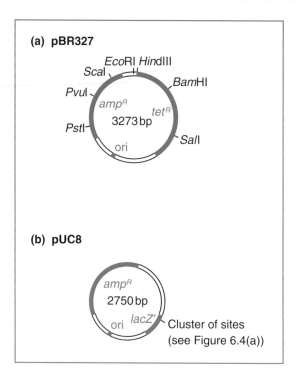

Figure 6.3 Two *E. coli* plasmid cloning vectors.

cases gene dosage becomes important, as the more copies there are of a cloned gene, the more likely it is that the effect of the cloned gene on the host cell will be detectable. pBR327, with its high copy number, is therefore a better choice than pBR322 for this kind of work.

(2) The deletion also destroys the conjugative ability of pBR322, making pBR327 a non-conjugative plasmid that cannot direct its own transfer to other *E. coli* cells. This is important for **biological containment**, averting the possibility of a recombinant pBR327 molecule escaping from the test tube and colonizing bacteria in the gut of a careless molecular biologist. In contrast, pBR322 could theoretically be passed to natural populations of *E. coli* by conjugation, though in fact pBR322 also has safeguards (though less sophisticated ones) to minimize the chances of this happening. pBR327 is, however, preferable if the cloned gene is potentially harmful should an accident occur.

pUC8 – a Lac selection plasmid

This vector was mentioned in Chapter 5 when identification of recombinants by insertional inactivation of the β-galactosidase gene was described (p. 94). pUC8 (Figure 6.3(b)) is derived from pBR322, although only the replication origin and the *amp*^{R} gene remain. The nucleotide sequence of the *amp*^{R} gene has been changed so that it no longer contains the unique restriction sites;

all these cloning sites are now clustered into a short segment of the *lacZ'* gene carried by pUC8.

pUC8 has three important advantages that have led to it becoming one of the most popular *E. coli* cloning vectors. The first of these is fortuitous: the manipulations involved in construction of pUC8 were accompanied by a chance mutation, within the origin of replication, that results in the plasmid having a copy number of 500–700 even before amplification. This has a significant effect on the yield of cloned DNA obtainable from *E. coli* cells transformed with recombinant pUC8 plasmids.

The second advantage is that identification of recombinant cells can be achieved by a single step process, by plating onto agar medium containing ampicillin plus X-gal (p. 95). With both pBR322 and pBR327, selection of recombinants is a two step procedure, requiring replica plating from one antibiotic medium to another (p. 93). A cloning experiment with pUC8 can therefore be carried out in half the time needed with pBR322 or pBR327.

The third advantage of pUC8 lies with the clustering of the restriction sites, which allows a DNA fragment with two different sticky ends (say *Eco*RI at one end and *Bam*HI at the other) to be cloned without resorting to additional manipulations such as linker attachment (Figure 6.4(a)). Other pUC vectors carry different combinations of restriction sites and provide even greater flexibility in the types of DNA fragment that can be cloned (Figure 6.4(b)). Furthermore, the restriction site clusters in these vectors are the same as the clusters in the equivalent M13mp series of vectors (p. 115). DNA cloned into a member of the pUC series can therefore be transferred directly to its M13mp counterpart, in which form it can be analysed by DNA sequencing or *in vitro* mutagenesis (Figure 6.4(c); see pp. 204 and 236 for descriptions of these procedures).

pGEM3Z – *in vitro* transcription of cloned DNA

pGEM3Z (Figure 6.5(a)) is very similar to a pUC vector: it carries the *amp*^{R} and *lacZ'* genes, the latter containing a cluster of restriction sites, and it is almost exactly the same size. The distinction is that pGEM3Z has two additional short pieces of DNA, each of which acts as the recognition site for attachment of an RNA polymerase enzyme. These two **promoter** sequences lie on either side of the cluster of restriction sites used for introduction of new DNA into the pGEM3Z molecule. This means that if a recombinant pGEM3Z molecule is mixed with purified RNA polymerase in the test tube, transcription occurs and RNA copies of the cloned fragment are synthesized (Figure 6.5(b)). The RNA that is produced could be used as a hybridization probe (p. 164), or might be required for experiments aimed at studying RNA processing (e.g. the removal of introns) or protein synthesis.

The promoters carried by pGEM3Z and other vectors of this type are not the standard sequences recognized by the *E. coli* RNA polymerase. Instead, one of the promoters is specific for the RNA polymerase coded by T7

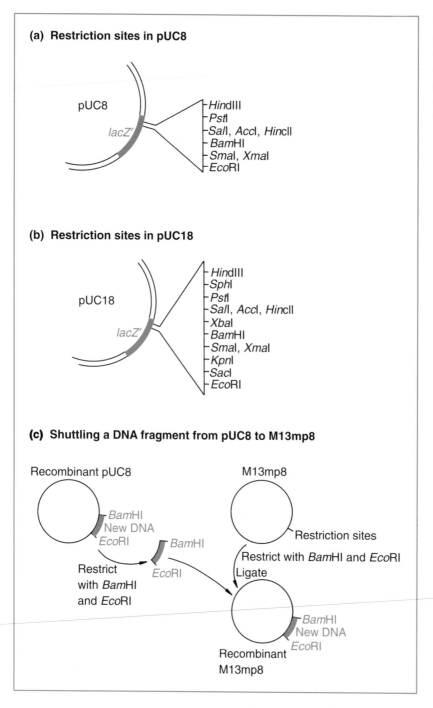

(a) Restriction sites in pUC8

pUC8

lacZ'

- HindIII
- PstI
- SalI, AccI, HincII
- BamHI
- SmaI, XmaI
- EcoRI

(b) Restriction sites in pUC18

pUC18

lacZ'

- HindIII
- SphI
- PstI
- SalI, AccI, HincII
- XbaI
- BamHI
- SmaI, XmaI
- KpnI
- SacI
- EcoRI

(c) Shuttling a DNA fragment from pUC8 to M13mp8

Recombinant pUC8

- BamHI
- New DNA
- EcoRI

Restrict with *Bam*HI and *Eco*RI

- BamHI
- EcoRI

M13mp8

Restriction sites

Restrict with *Bam*HI and *Eco*RI

Ligate

Recombinant M13mp8

- BamHI
- New DNA
- EcoRI

Figure 6.4 The pUC plasmids. (a) The restriction site cluster in the *lacZ'* gene of pUC8. (b) The restriction site cluster in pUC18. (c) Shuttling a DNA fragment from pUC8 to M13mp8.

bacteriophage and the other for the RNA polymerase of SP6 phage. These RNA polymerases are synthesized during infection of *E. coli* with one or other of the phages and are responsible for transcribing the phage genes. They are chosen for use in *in vitro* transcription as they are very active

Figure 6.5 pGEM3Z. (a) Map of the vector. (b) *In vitro* RNA synthesis. R = cluster of restriction sites for *Eco*RI, *Sac*I, *Kpn*I, *Ava*I, *Sma*I, *Bam*HI, *Xba*I, *Sal*I, *Acc*I, *Hinc*II, *Pst*I, *Sph*I and *Hind*III.

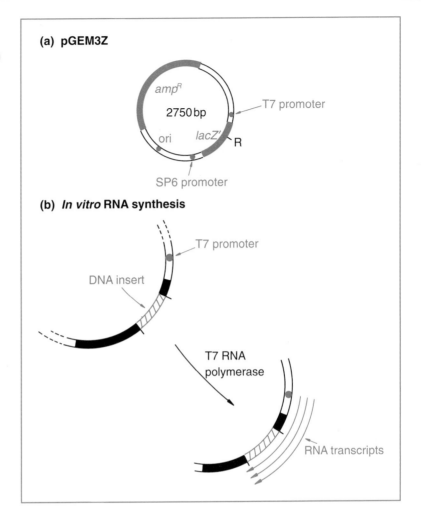

enzymes (remember that the entire lytic infection cycle takes only 20 min (p. 20), so the phage genes must be transcribed very quickly) and are able to synthesize 1–2 μg of RNA per minute, substantially more than can be produced by the standard *E. coli* enzyme.

6.2 Cloning vectors based on M13 bacteriophage

The most essential requirement for any cloning vector is that it has a means of replicating in the host cell. For plasmid vectors this requirement is easy to satisfy, as relatively short DNA sequences are able to act as plasmid origins of replication, and most, if not all, of the enzymes needed for replication

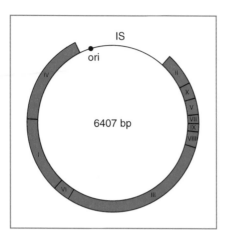

Figure 6.6 The M13 genome, showing the positions of genes I to X.

are provided by the host cell. Elaborate manipulations, such as those that resulted in pBR322 (Figure 6.2(a)), are therefore possible so long as the final construction has an intact, functional replication origin.

With bacteriophages such as M13 the situation as regards replication is more complex. Phage DNA molecules generally carry several genes that are essential for replication, including genes coding for components of the phage protein coat and phage-specific DNA replicative enzymes. Alteration or deletion of any of these genes will impair or destroy the replicative ability of the resulting molecule. There is therefore much less freedom to modify phage DNA molecules, and generally phage cloning vectors are only slightly different from the parent molecule.

The problems in constructing a phage cloning vector are illustrated by considering M13. The normal M13 genome is 6.4 kb in length, but most of this is taken up by ten closely packed genes (Figure 6.6), each essential for the replication of the phage. There is only a single, 507 nucleotide intergenic sequence into which new DNA could be inserted without disrupting one of these genes, and in fact this region includes the replication origin which must itself remain intact. Clearly there is only limited scope for modifying the M13 genome.

Nevertheless, it will be remembered that the great attraction of M13 is the opportunity it offers of obtaining single-stranded versions of cloned DNA (p. 25). This feature has acted as a stimulus for the development of M13 cloning vectors.

6.2.1   Development of the cloning vector M13mp2

The first step in construction of an M13 cloning vector was to introduce the *lacZ'* gene into the intergenic sequence. This gave rise to M13mp1, which forms blue plaques on X-gal agar (Figure 6.7(a)).

M13mp1 does not possess any unique restriction sites in the *lacZ'* gene. It does, however, contain the hexanucleotide GGATTC near the start of the

Figure 6.7 Construction of (a) M13mp1, and (b) M13mp2 from the wild-type M13 genome.

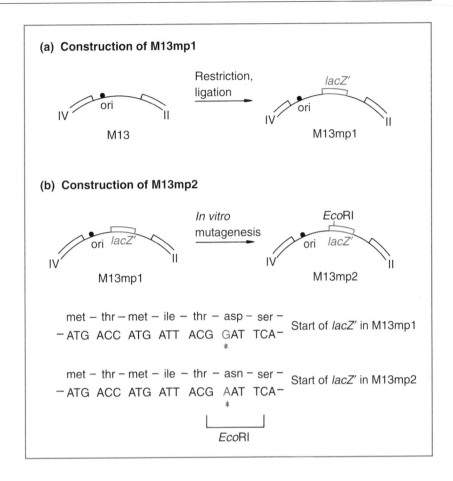

gene. A single nucleotide change would make this GAATTC, which is an *Eco*RI site. This alteration was carried out using *in vitro* mutagenesis (p. 236), resulting in M13mp2 (Figure 6.7(b)). M13mp2 has a slightly altered *lacZ'* gene (the sixth codon now specifies asparagine instead of aspartic acid), but the β-galactosidase enzyme produced by cells infected with M13mp2 is still perfectly functional.

M13mp2 is the simplest M13 cloning vector. DNA fragments with *Eco*RI sticky ends can be inserted into the cloning site, and recombinants are distinguished as clear plaques on X-gal agar.

6.2.2 M13mp7 – symmetrical cloning sites

The next step in the development of M13 vectors was to introduce additional restriction sites into the *lacZ'* gene. This was achieved by synthesizing in the test tube a short oligonucleotide, called a **polylinker**, that consists of a series of restriction sites and has *Eco*RI sticky ends (Figure 6.8(a)). This polylinker was inserted into the *Eco*RI site of M13mp2, to give M13mp7 (Figure 6.8(b)), a more complex vector with four possible cloning sites

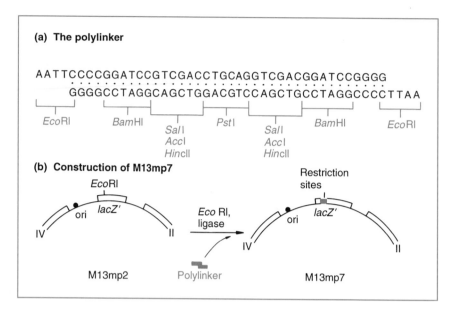

Figure 6.8 Construction of M13mp7: (a) the polylinker, and (b) its insertion into the *Eco*RI site of M13mp2. Note that the *Sal*I restriction sites are also recognized by *Acc*I and *Hinc*II.

(*Eco*RI, *Bam*HI, *Sal*I and *Pst*I). The polylinker is designed so that it does not totally disrupt the *lacZ'* gene; a reading frame is maintained throughout the polylinker, and a functional, though altered, β-galactosidase enzyme is still produced.

When M13mp7 is digested with either *Eco*RI, *Bam*HI or *Sal*I, a part or all of the polylinker is excised (Figure 6.9(a)). On ligation, in the presence of new DNA, one of three events may occur (Figure 6.9(b)).

(1) New DNA is inserted.
(2) The polylinker is reinserted.
(3) The vector self-ligates without insertion.

Insertion of new DNA almost invariably prevents β-galactosidase production, so recombinant plaques are clear on X-gal agar (Figure 6.9(c)). Alternatively, if the polylinker is reinserted, and the original M13mp7 reformed, then blue plaques result. But what if the vector self-ligates, with neither new DNA nor the polylinker inserted? Once again the design of the polylinker comes into play. Whichever restriction site is used, self-ligation results in a functional *lacZ'* gene (Figure 6.9(c)), giving blue plaques. Selection is therefore unequivocal: only recombinant M13mp7 phage give rise to clear plaques.

A big advantage of M13mp7, with its symmetrical cloning sites, is that DNA inserted into either the *Bam*HI, *Sal*I or *Pst*I sites can be excised from the recombinant molecule using *Eco*RI (Figure 6.10). Very few vectors allow cloned DNA to be recovered so easily.

Figure 6.9 Cloning with
M13mp7 (see text for details).

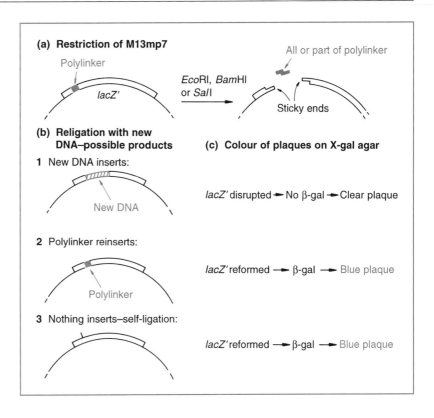

(a) Restriction of M13mp7

Polylinker

All or part of polylinker

lacZ'

EcoRI, BamHI
or SalI

Sticky ends

**(b) Religation with new
    DNA–possible products**

1 New DNA inserts:

New DNA

(c) Colour of plaques on X-gal agar

lacZ' disrupted → No β-gal → Clear plaque

2 Polylinker reinserts:

Polylinker

lacZ' reformed → β-gal → Blue plaque

3 Nothing inserts–self-ligation:

lacZ' reformed → β-gal → Blue plaque

6.2.3

More complex M13 vectors

The most sophisticated M13 vectors have more complex polylinkers inserted into the *lacZ'* gene. An example is M13mp8 (Figure 6.11(a)), which is the counterpart of the plasmid pUC8 (p. 108). As with the plasmid vector, one advantage of M13mp8 is its ability to take DNA fragments with two different sticky ends.

A second feature is provided by the sister vector M13mp9 (Figure 6.11(b)), which has the same polylinker but in the reverse orientation. A DNA fragment cloned into M13mp8, if excised by double restriction, and then inserted into M13mp9, will now itself be in the reverse orientation (Figure 6.11(c)). This is important in DNA sequencing (p. 204), during which the nucleotide sequence is read from one end of the polylinker into the inserted DNA fragment (Figure 6.11 (d)). Only about 600 nucleotides can be read from one sequencing experiment; if the inserted DNA is longer than this, one end of the fragment will not be sequenced. One answer is to turn the fragment around, by excising it and reinserting into the sister vector. A DNA sequencing experiment with this new clone allows the nucleotide sequence at the other end of the fragment to be determined.

Other M13 vector pairs are also available. M13mp10/11 and M13mp18/19 are similar to M13mp8/9, but have different polylinkers and therefore different restriction sites.

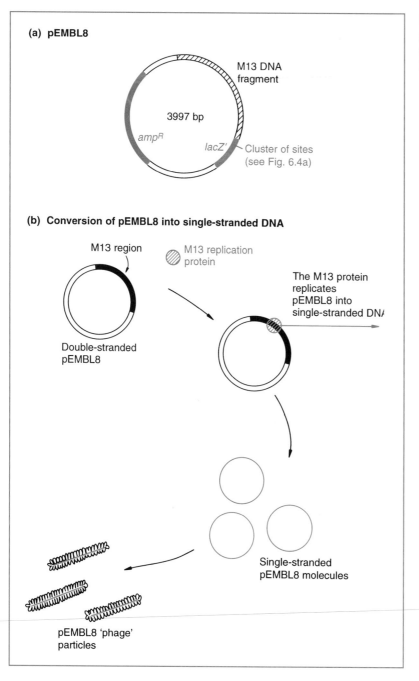

Figure 6.12 pEMBL8: a hybrid plasmid–M13 vector that can be converted into single-stranded DNA.

derived from pUC8, has the polylinker cloning sites within the *lacZ'* gene, so recombinant plaques can be identified in the standard way on agar containing X-gal. With pEMBL8, single-stranded versions of cloned DNA fragments up to 10 kb in length can be obtained, greatly extending the range of the M13 cloning system.

6.3

Cloning vectors based on λ bacteriophage

Two problems had to be solved before λ-based cloning vectors could be developed:

(1) The λ DNA molecule can be increased in size by only about 5%, representing the addition of only 3 kb of new DNA. If the total size of the molecule is more than 52 kb, then it cannot be packaged into the λ head structure and infective phage particles are not formed. This severely limits the size of a DNA fragment that can be inserted into an unmodified λ vector (Figure 6.13(a)).

(2) The λ genome is so large that it has more than one recognition sequence for virtually every restriction endonuclease. Restriction cannot be used to cleave the normal λ molecule in a way that will allow insertion of new DNA, because the molecule would be cut into several small fragments that would be very unlikely to re-form a viable λ genome on re-ligation (Figure 6.13(b)).

Figure 6.13 The two problems that had to be solved before λ cloning vectors could be developed. (a) The size limitation placed on the λ genome by the need to package it into the phage head. (b) λ DNA has multiple recognition sites for almost all restriction endonucleases.

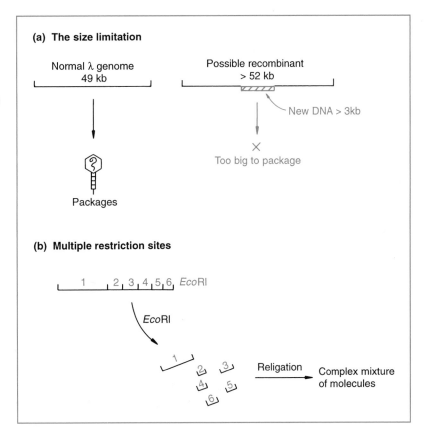

(a) The size limitation

Normal λ genome
49 kb

Possible recombinant
> 52 kb

New DNA > 3kb

Packages

✕
Too big to package

(b) Multiple restriction sites

1   2, 3, 4, 5, 6, *Eco*RI

*Eco*RI

1
2 3
4 5
6

Religation

Complex mixture of molecules

In view of these difficulties it is perhaps surprising that a wide variety of λ cloning vectors have been developed, their primary use being to clone large pieces of DNA, from 5–25 kb, much too big to be handled by plasmid or M13 vectors.

6.3.1

Segments of the λ genome can be deleted without impairing viability

The way forward for the development of λ cloning vectors was provided by the discovery that a large segment in the central region of the λ DNA molecule can be removed without affecting the ability of the phage to infect *E. coli* cells. Removal of all or part of this non-essential region, between positions 20 and 35 on the map shown in Figure 2.9, decreases the size of the resulting λ molecule by up to 15 kb. This means that as much as 18 kb of new DNA can now be added before the cut-off point for packaging is reached (Figure 6.14).

The 'non-essential' region in fact contains most of the genes involved in integration and excision of the λ prophage from the *E. coli* chromosome. A deleted λ genome is therefore non-lysogenic and can follow only the lytic infection cycle. This in itself is desirable for a cloning vector as it means induction is not needed before plaques are formed (p. 46).

6.3.2

Natural selection can be used to isolate modified λ that lack certain restriction sites

Even a deleted λ genome, with the non-essential region removed, has multiple recognition sites for most restriction endonucleases. This is a problem that is often encountered when a new vector is being developed. If just one or two sites need to be removed then the technique of *in vitro* mutagenesis (p. 236) can be used. For example, an *Eco*RI site, GAATTC, could be changed to GGATTC, which is not recognized by the enzyme. However, *in vitro* mutagenesis was in its infancy when the first λ vectors were under development, and even today would not be an efficient means of changing more than a few sites in a single molecule.

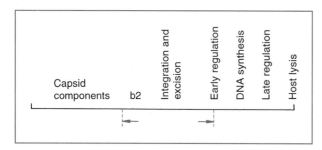

Figure 6.14 The λ genetic map, showing the position of the non-essential region that can be deleted without affecting the ability of the phage to follow the lytic infection cycle.

Figure 6.15 Using natural selection to isolate λ phage lacking *Eco*RI restriction sites.

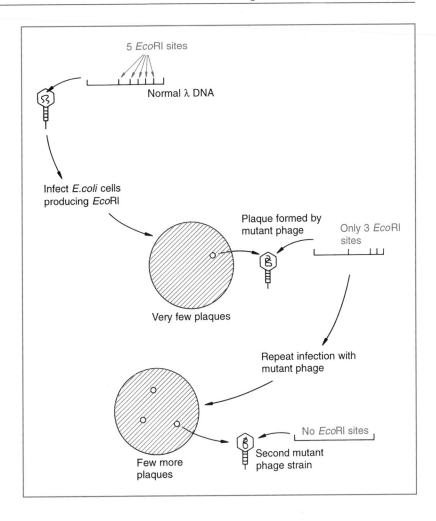

Instead, natural selection was used to provide strains of λ that lack the unwanted restriction sites. Natural selection can be brought into play by using as a host an *E. coli* strain that produces *Eco*RI. Most λ DNA molecules that invade the cell are destroyed by this restriction endonuclease, but a few survive and produce plaques. These are mutant phage, from which one or more *Eco*RI sites have been lost spontaneously (Figure 6.15). Several cycles of infection will eventually result in λ molecules that lack all or most of the *Eco*RI sites.

6.3.3 Insertion and replacement vectors

Once the problems posed by packaging constraints and by the multiple restriction sites had been solved, the way was open for the development of different types of λ-based cloning vectors. The first two classes of vector to be produced were λ **insertion** and λ **replacement** (or substitution) vectors.

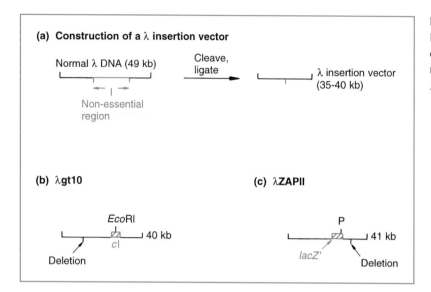

(a) Construction of a λ insertion vector

Normal λ DNA (49 kb)

Cleave, ligate

λ insertion vector (35-40 kb)

Non-essential region

(b) λgt10

EcoRI

40 kb

cI

Deletion

(c) λZAPII

P

41 kb

lacZ'

Deletion

Figure 6.16 λ insertion vectors. P = polylinker in the *lacZ'* gene of λZAPII, containing unique restriction sites for *Sac*I, *Not*I, *Xba*I, *Spe*I, *Eco*RI and *Xho*I.

Insertion vectors

With an insertion vector (Figure 6.16(a)), a large segment of the non-essential region has been deleted, and the two arms ligated together. An insertion vector possesses at least one unique restriction site into which new DNA can be inserted. The size of the DNA fragment that an individual vector can carry depends of course on the extent to which the non-essential region has been deleted. Two popular insertion vectors are:

- **λgt10** (Figure 6.16(b)), which can carry up to 8 kb of new DNA, inserted into a unique *Eco*RI site located in the *c*I gene. Insertional inactivation of this gene means that recombinants are distinguished as clear rather than turbid plaques (p. 100).
- **λZAPII** (Figure 6.16(c)), with which insertion of up to 10 kb DNA into any of six restriction sites within a polylinker inactivates the *lacZ'* gene carried by the vector. Recombinants give clear rather than blue plaques on X-gal agar.

Replacement vectors

A λ replacement vector has two recognition sites for the restriction endonuclease used for cloning. These sites flank a segment of DNA that is replaced by the DNA to be cloned (Figure 6.17(a)). Often the replaceable fragment (or '**stuffer fragment**' in cloning jargon) carries additional restriction sites that can be used to cut it up into small pieces, so that its own reinsertion during a cloning experiment is very unlikely. Replacement vectors are generally designed to carry larger pieces of DNA than inser-

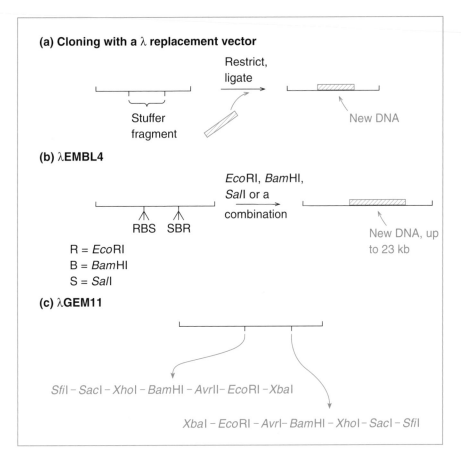

Figure 6.17 λ replacement vectors. (a) Cloning with a λ replacement vector. (b) Cloning with λEMBL4. (c) The structure of λGEM11, showing the order of restriction sites in the two polylinkers.

tion vectors can handle. Recombinant selection is often on the basis of size, with non-recombinant vectors being too small to be packaged into λ phage heads (p. 101).

Two popular replacement vectors are:

- **λEMBL4** (Figure 6.17(b)) can carry up to 20 kb of inserted DNA by replacing a segment flanked by pairs of *Eco*RI, *Bam*HI and *Sal*I sites (any of these three restriction endonucleases can be used to remove the stuffer fragment, so DNA fragments with a variety of sticky ends can be cloned). Recombinant selection with λEMBL4 can be on the basis of size, or can utilize the Spi phenotype (p. 101).
- **λGEM11** and **λGEM12** (Figure 6.17(c)), each of which has a capacity of 23 kb (near the theoretical maximum) with the stuffer fragment flanked by polylinkers that contain seven different restriction sites. The polylinkers are slightly different in the two vectors, but in both cases the outermost sites are for the restriction enzyme *Sfi*I, which recognizes the sequence GGCCNNNNNGGCC. This sequence is very rare and

unlikely to be present in the DNA fragment that has been cloned. Restriction of the recombinant vector with *Sfi*I can therefore be used to cut out the cloned fragment, with a high likelihood that the fragment will be recovered intact. As with λEMBL4, size or the Spi phenotype is used to select recombinants.

6.3.4

Cloning experiments with λ insertion or replacement vectors

A cloning experiment with a λ vector can proceed along the same lines as with a plasmid vector – the λ molecules are restricted, new DNA is added, the mixture is ligated, and the resulting molecules used to transfect a competent *E. coli* host (Figure 6.18(a)). This type of experiment requires that the vector be in its circular form, with the *cos* sites hydrogen bonded to each other.

Although quite satisfactory for many purposes, a procedure based on transfection is not particularly efficient. A greater number of recombinants

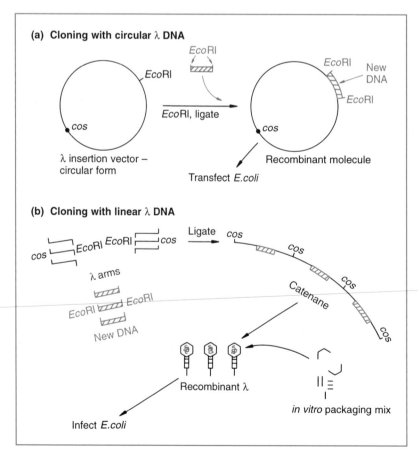

Figure 6.18 Different strategies for cloning with a λ vector. (a) Using the circular form of λ as a plasmid. (b) Using left and right arms of the λ genome, plus *in vitro* packaging, to achieve a greater number of recombinant plaques.

will be obtained if one or two refinements are introduced. The first is to use the linear form of the vector. When the linear form of the vector is digested with the relevant restriction endonuclease, the left and right arms are released as separate fragments. A recombinant molecule can be constructed by mixing together the DNA to be cloned with the vector arms (Figure 6.18(b)). Ligation results in several molecular arrangements, including catenanes comprising left arm–DNA–right arm repeated many times (Figure 6.18(b)). If the inserted DNA is the correct size then the *cos* sites that separate these structures will be the right distance apart for *in vitro* packaging (p. 91). Recombinant phage are therefore produced in the test tube and can be used to infect an *E. coli* culture. This strategy, in particular the use of *in vitro* packaging, results in a large number of recombinant plaques.

<table>
<tr><td>6.3.5</td><td>

Very large DNA fragments can be cloned using a cosmid

</td></tr>
</table>

The final and most sophisticated type of λ-based vector is the **cosmid**. Cosmids are hybrids between a phage DNA molecule and a bacterial plasmid, and their design centres on the fact that the enzymes that package the λ DNA molecule into the phage protein coat need only the *cos* sites in order to function (p. 24). The *in vitro* packaging reaction works not only with λ genomes, but also with any molecule that carries *cos* sites separated by 37–52 kb of DNA.

A cosmid is basically a plasmid that carries a *cos* site (Figure 6.19(a)). It also needs a selectable marker, such as the ampicillin resistance gene, and a plasmid origin of replication, as cosmids lack all the λ genes and so do not produce plaques. Instead colonies are formed on selective media, just as with a plasmid vector.

A cloning experiment with a cosmid is carried out as follows (Figure 6.19(b)). The cosmid is opened at its unique restriction site and new DNA fragments inserted. These fragments are usually produced by partial digestion with a restriction endonuclease, as total digestion almost invariably results in fragments that are too small to be cloned with a cosmid. Ligation is carried out so that catenanes are formed. Providing the inserted DNA is the right size, *in vitro* packaging cleaves the *cos* sites and places the recombinant cosmids in mature phage particles. These λ phage are then used to infect an *E. coli* culture, though of course plaques are not formed. Instead, infected cells are plated onto a selective medium and antibiotic-resistant colonies are grown. All colonies are recombinants as non-recombinant linear cosmids are too small to be packaged into λ heads.

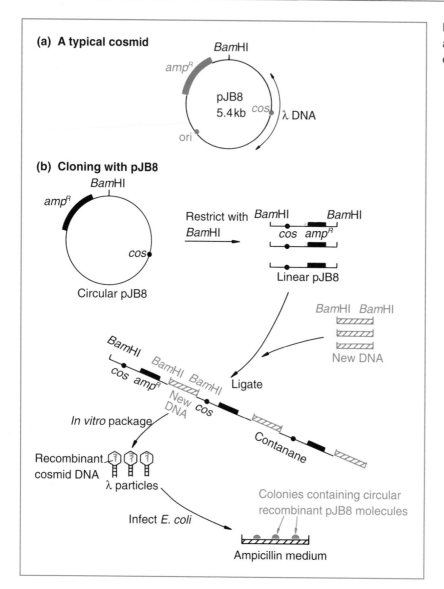

Figure 6.19 A typical cosmid and the way it is used to clone large fragments of DNA.

6.4

λ and other high capacity vectors enable genomic libraries to be constructed

The main use of all λ-based vectors is to clone DNA fragments that are too large to be handled by plasmid or M13 vectors. A replacement vector such as λEMBL4 can carry up to 20kb of new DNA, and some cosmids can manage fragments up to 40kb. This compares with a maximum insert size of about 8kb for most plasmids and less than 3kb for M13 vectors.

The ability to clone such large DNA fragments means that **genomic libraries** can be generated. A genomic library is a set of recombinant clones that contain all of the DNA present in an individual organism. An *E. coli* genomic library, for example, contains all the *E. coli* genes, so any desired gene can be withdrawn from the library and studied. Genomic libraries can be retained for many years, and propagated so that copies can be sent from one research group to another.

The big question is, how many clones are needed for a genomic library? The answer can be calculated with the formula:

$$N = \frac{\ln(1-P)}{\ln\left(1 - \dfrac{a}{b}\right)}$$

where N is the number of clones that are required, P is probability that any given gene will be present, a is the average size of the DNA fragments inserted into the vector, and b is the total size of the genome. Table 6.1 shows the number of clones needed for genomic libraries of a variety of organisms, constructed using a λ replacement vector or a cosmid.

It is by no means impossible to obtain several hundred thousand clones, and the methods used to identify a clone carrying a desired gene (Chapter 8) can be adapted to handle such large numbers, so genomic libraries of these sizes are by no means unreasonable. However, ways of reducing the number of clones needed for a genomic library are continually being sought.

One solution is to develop new cloning vectors able to handle larger DNA inserts. During the last few years, progress in this area has centred on **bacterial artificial chromosomes (BACs)**, which are novel vectors based on the F

Table 6.1 Number of clones needed for genomic libraries of a variety of organisms.

Species	Genome size (bp)	Number of clones[a]	
		17 kb fragments[b]	35 kb fragments[c]
E. coli	4.6×10^6	820	410
Saccharomyces cerevisiae	1.8×10^7	3225	1500
Drosophila melanogaster	1.2×10^8	21500	10000
Rice	5.7×10^8	100000	49000
Human	3.2×10^9	564000	274000
Frog	2.3×10^{10}	4053000	1969000

[a] Calculated for a probability (*P*) of 95% that any particular gene will be present in the library.
[b] Fragments suitable for a replacement vector such as λEMBL4.
[c] Fragments suitable for a cosmid.

plasmid (p. 18). The F plasmid is relatively large and vectors derived from it have a higher capacity than normal plasmid vectors. BACs can handle DNA inserts up to 300 kb in size, reducing the size of the human genomic library to just 30 000 clones. Other high capacity vectors have been constructed from bacteriophage **P1**, which has the advantage over λ of being able to squeeze 110 kb of DNA into its capsid structure. Cosmid-type vectors based on P1 have been designed and used to clone DNA fragments ranging in size from 75 to 100 kb. Vectors that combine the features of P1 vectors and BACs, called **P1-derived artificial chromosomes (PACs)**, also have a capacity of up to 300 kb.

6.5 Vectors for other bacteria

Cloning vectors have also been developed for several other species of bacteria, including *Streptomyces*, *Bacillus* and *Pseudomonas*. Some of these vectors are based on plasmids specific to the host organism, and some on **broad host range plasmids** able to replicate in a variety of bacterial hosts. A few are derived from bacteriophages specific to these organisms. Most of these vectors are very similar to *E. coli* vehicles in terms of general purposes and uses.

Further reading

Bolivar, F., Rodriguez, R.L., Green, P.J. *et al.* (1977) Construction and characterization of new cloning vectors. II. A multipurpose cloning system. *Gene*, **2**, 95–113. [pBR322.]

Collins, J. & Hohn, B. (1978) Cosmids: a type of plasmid gene cloning vector that is packageable *in vitro* in bacteriophage heads. *Proceedings of the National Academy of Sciences of the USA*, **75**, 4242–6.

Dente, L., Cesareni, G. & Cortese, R. (1983) pEMBL: a new family of single-stranded plasmids. *Nucleic Acids Research*, **11**, 1645–55.

Frischauf, A.-M., Lehrach, H., Poustka, A. & Murray, N. (1983) Lambda replacement vectors carrying polylinker sequences. *Journal of Molecular Biology*, **170**, 827–42. [The λEMBL vectors.]

Ioannou, P.A., Amemiya, C.T., Garnes, J. *et al.* (1994) P1-derived vector for the propagation of large human DNA fragments. *Nature Genetics*, **6**, 84–9.

Melton, D.A., Krieg, P.A., Rebagliati, M.R., Maniatis, T., Zinn, K. & Green, M.R. (1984) Efficient *in vitro* synthesis of biologically active RNA and RNA hybridization probes from plasmids containing a bacteriophage SP6 promoter. *Nucleic Acids Research*, **12**, 7035–56. [RNA synthesis from DNA cloned in a plasmid such as pGEM3Z.]

Sanger, F., Coulson, A.R., Barrell, B.G., Smith, A.J.M. & Roe, B.A. (1980) Cloning in single-stranded bacteriophage as an aid to rapid DNA sequencing. *Journal of Molecular Biology*, **143**, 161–78. [M13 vectors.]

Shiyuza, H., Birren, B., Kim, U.J. *et al.* (1992) Cloning and stable maintenance of 300 kilobase-pair fragments of human DNA in *Escherichia coli* using an F-factor-based vector. *Proceedings of the National Academy of Sciences of the USA*, **89**, 8794–7. [The first description of a BAC.]

Sternberg, N. (1990) Bacteriophage P1 cloning system for the isolation, amplification, and recovery of DNA fragments as large as 100 kilobase pairs. *Proceedings of the National Academy of Sciences of the USA*, **87**, 103–7.

Yanisch-Perron, C., Vieira, J. & Messing, J. (1985) Improved M13 phage cloning vectors and host strains: nucleotide sequences of the M13mp18 and pUC19 vectors. *Gene*, **33**, 103–19.

Chapter 7

Cloning Vectors for Eukaryotes

Most cloning experiments are carried out with *E. coli* as the host, and the widest variety of cloning vectors are available for this organism. *E. coli* is particularly popular when the aim of the cloning experiment is to study the basic features of molecular biology such as gene structure and function. However, under some circumstances it may be desirable to use a different host for a gene cloning experiment. This is especially true in biotechnology (Chapter 13), where the aim may not be to study a gene, but to use cloning to control or improve synthesis of an important metabolic product (e.g. a hormone such as insulin), or to change the properties of the organism (e.g. to introduce herbicide resistance into a crop plant). We must therefore consider cloning vectors for organisms other than *E. coli*.

7.1 Vectors for yeast and other fungi

The yeast *Saccharomyces cerevisiae* is one of the most important organisms in biotechnology. As well as its role in brewing and breadmaking, yeast has been used as a host organism for the production of important pharmaceuticals from cloned genes (p. 286). Development of cloning vectors for yeast has been stimulated greatly by the discovery of a plasmid that is present in most strains of *S. cerevisiae* (Figure 7.1). The 2 μm plasmid, as it is called, is one of only a very limited number of plasmids found in eukaryotic cells.

7.1.1 Selectable markers for the 2 μm plasmid

The 2 μm plasmid is in fact an excellent basis for a cloning vector. It is 6 kb in size, which is ideal for a vector, and exists in the yeast cell at a copy number of between 70 and 200. Replication makes use of a plasmid origin, several enzymes provided by the host cell, and the proteins coded by the *REP1* and *REP2* genes carried by the plasmid.

Figure 7.1 The yeast 2 μm plasmid. *REP1* and *REP2* are involved in replication of the plasmid, and *FLP* codes for a protein that can convert the A form of the plasmid (shown here) to the B form, in which the gene order has been rearranged by intramolecular recombination. The function of *D* is not exactly known.

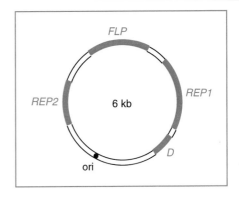

However, all is not perfectly straightforward in using the 2 μm plasmid as a cloning vector. First, there is the question of a selectable marker. Some yeast cloning vectors carry genes conferring resistance to inhibitors such as methotrexate and copper, but most of the popular yeast vectors make use of a radically different type of selection system. In practice a normal yeast gene is used, generally one that codes for an enzyme involved in amino acid biosynthesis. An example is the gene *LEU2*, which codes for β-isopropyl malate dehydrogenase, one of the enzymes involved in the conversion of pyruvic acid to leucine.

In order to use *LEU2* as a selectable marker, a special kind of host organism is needed. The host must be an **auxotrophic** mutant that has a non-functional *LEU2* gene. Such a *leu2⁻* yeast is unable to synthesize leucine and can survive only if this amino acid is supplied as a nutrient in the growth medium (Figure 7.2(a)). Selection is possible because transformants contain a plasmid-borne copy of the *LEU2* gene, and so are able to grow in the absence of the amino acid. In a cloning experiment, cells are plated out onto **minimal medium** (which contains no added amino acids). Only transformed cells are able to survive and form colonies (Figure 7.2(b)).

7.1.2   Vectors based on the 2 μm plasmid – yeast episomal plasmids

Vectors derived from the 2 μm plasmid are called **yeast episomal plasmids (YEps)**. Some YEps contain the entire 2 μm plasmid, others include just the 2 μm origin of replication. An example of the latter type is YEp13 (Figure 7.3).

YEp13 illustrates several general features of yeast cloning vectors. First, it is a **shuttle vector**. As well as the 2 μm origin of replication and the selectable *LEU2* gene, YEp13 also includes the entire pBR322 sequence, and can therefore replicate and be selected for in both yeast and *E. coli*. There are several lines of reasoning behind the use of shuttle vectors. One is that it may be difficult to recover the recombinant DNA molecule from a transformed yeast colony. This is not such a problem with YEps, which are present in yeast

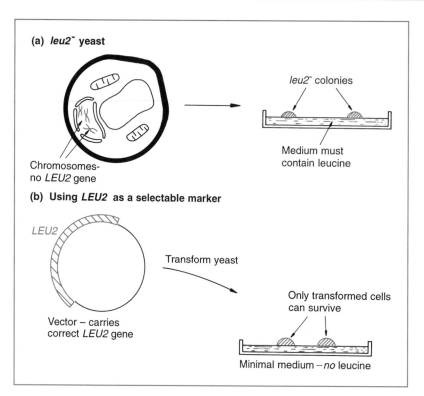

(a) *leu2⁻* **yeast**

leu2⁻ colonies

Chromosomes-
no *LEU2* gene

Medium must
contain leucine

(b) Using LEU2 as a selectable marker

LEU2

Transform yeast

Only transformed cells
can survive

Vector – carries
correct *LEU2* gene

Minimal medium −*no* leucine

Figure 7.2 Using the *LEU2* gene as a selectable marker in a yeast cloning experiment.

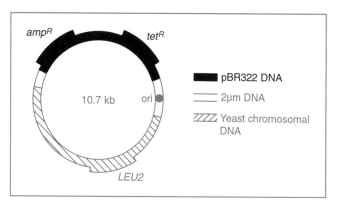

*amp*ᴿ

*tet*ᴿ

10.7 kb

ori

■ pBR322 DNA

— 2μm DNA

▨ Yeast chromosomal DNA

LEU2

Figure 7.3 A yeast episomal plasmid, YEp13.

cells primarily as plasmids, but with other yeast vectors, which may integrate into one of the yeast chromosomes (p. 133), purification may be impossible. This is a disadvantage because in many cloning experiments purification of recombinant DNA is essential in order for the correct construct to be identified by, for example, DNA sequencing.

The standard procedure when cloning in yeast is therefore to perform the initial cloning experiment with *E. coli*, and to select recombinants in this organism. Recombinant plasmids can then be purified, characterized, and the correct molecule introduced into yeast (Figure 7.4).

Figure 7.4 Cloning with an *E. coli*–yeast shuttle vector such as YEp13.

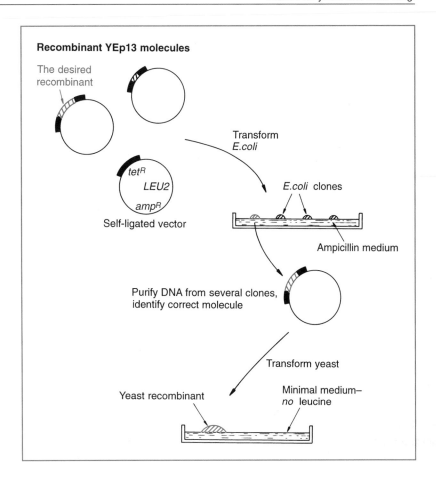

Recombinant YEp13 molecules

The desired recombinant

Transform *E. coli*

tet^R
LEU2
amp^R

Self-ligated vector

E.coli clones

Ampicillin medium

Purify DNA from several clones, identify correct molecule

Transform yeast

Yeast recombinant

Minimal medium– *no* leucine

7.1.3 A YEp may insert into yeast chromosomal DNA

The word 'episomal' indicates that a YEp can replicate as an independent plasmid, but also implies that integration into one of the yeast chromosomes can occur (see the definition of 'episome' on p. 16). Integration occurs because the gene carried on the vector as a selectable marker is very closely homologous to the mutant version of the gene present in the yeast chromosomal DNA. With YEp13, for example, recombination can occur between the plasmid *LEU2* gene and the yeast mutant *LEU2* gene, resulting in insertion of the entire plasmid into one of the yeast chromosomes (Figure 7.5). The plasmid may remain integrated, or a later recombination event may result in it being excised again.

7.1.4 Other types of yeast cloning vector

In addition to YEps, there are several other types of cloning vector for use with *S. cerevisiae*. Two important ones are as follows.

(1) **Yeast integrative plasmids (YIps)** are basically bacterial plasmids carrying a yeast gene. An example is YIp5, which is pBR322 with an

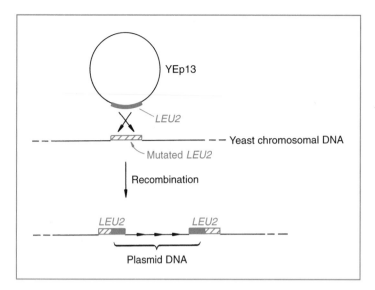

YEp13

LEU2

Yeast chromosomal DNA

Mutated LEU2

Recombination

LEU2

LEU2

Plasmid DNA

Figure 7.5 Recombination between plasmid and chromosomal *LEU2* genes can integrate YEp13 into yeast chromosomal DNA. After integration there are two copies of the *LEU2* gene; usually one is functional, and the other mutated.

inserted *URA3* gene (Figure 7.6(a)). This gene codes for orotidine-5′-phosphate decarboxylase (an enzyme that catalyses one of the steps in the biosynthesis pathway for pyrimidine nucleotides) and is used as a selectable marker in exactly the same way as *LEU2*. A YIp cannot replicate as a plasmid as it does not contain any parts of the 2 μm plasmid, and instead depends for its survival on integration into yeast chromosomal DNA. Integration occurs just as described for a YEp (Figure 7.5).

(2) **Yeast replicative plasmids (YRps)** are able to multiply as independent plasmids because they carry a chromosomal DNA sequence that includes an origin of replication. Replication origins are known to be located very close to several yeast genes, including one or two which can be used as selectable markers. YRp7 (Figure 7.6(b)) is an example of a replicative plasmid. It is made up of pBR322 plus the yeast gene *TRP1*. This gene, which is involved in tryptophan biosynthesis, is located adjacent to a chromosomal origin of replication. The yeast DNA fragment present in YRp7 contains both *TRP1* and the origin.

Three factors come into play when deciding which type of yeast vector is most suitable for a particular cloning experiment. The first of these is **transformation frequency**, a measure of the number of transformants that can be obtained per microgram of plasmid DNA. A high transformation frequency is necessary if a large number of recombinants are needed, or if the starting DNA is in short supply. YEps have the highest transformation frequency, providing between 10000 and 100000 transformed cells per μg. YRps are also quite productive, giving between 1000 and 10000 transformants per μg, but a YIp yields less than 1000 transformants per μg, and only 1–10 unless special procedures are used. The low transformation frequency of a YIp reflects the

Figure 7.6 A YIp and a YRp.

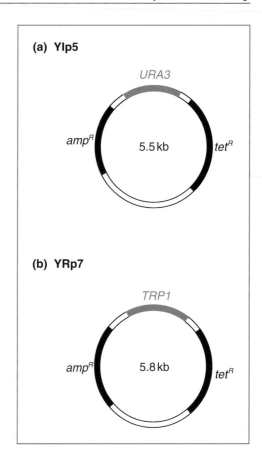

(a) YIp5

URA3

amp^R

5.5 kb

tet^R

(b) YRp7

TRP1

amp^R

5.8 kb

tet^R

fact that the rather rare chromosomal integration event is necessary before the vector can be retained in a yeast cell.

Furthermore, YEps and YRps also have the highest copy numbers: 20–50 and 5–100 respectively. In contrast, a YIp is usually present at just one copy per cell. These figures are important if the objective is to obtain protein from the cloned gene, as the more copies there are of the gene the greater the expected yield of the protein product.

So why would one ever wish to use a YIp? The answer is because YIps produce very stable recombinants, as loss of a YIp that has become integrated into a chromosome occurs at only a very low frequency. On the other hand, YRp recombinants are extremely unstable, the plasmids tending to congregate in the mother cell when a daughter cell buds off, so the daughter cell is non-recombinant; YEp recombinants suffer from similar problems, though an improved understanding of the biology of the 2 μm plasmid has enabled more stable YEps to be developed in recent years. Nevertheless, a YIp is the vector of choice if the needs of the experiment dictate that the recombinant yeast cells must retain the cloned gene for long periods in culture.

7.1.5

Artificial chromosomes can be used to clone large pieces of DNA in yeast

The final type of yeast cloning vector to consider is the **yeast artificial chromosome (YAC)**, a totally new approach to gene cloning. The development of YACs has been a spin-off from fundamental research into the structure of eukaryotic chromosomes, work that has identified the key components of a chromosome as being (Figure 7.7):

(1) The centromere, which is required for the chromosome to be distributed correctly to daughter cells during cell division.

(2) Two telomeres, the structures at the ends of a chromosome, which are needed in order for the ends to be replicated correctly and which also prevent the chromosome from being nibbled away by exonucleases.

(3) The origins of replication, which are the positions along the chromosome at which DNA replication initiates, similar to the origin of replication of a plasmid.

Once chromosome structure had been defined in this way the possibility arose that the individual components might be isolated by recombinant DNA techniques and then joined together again in the test tube, creating an artificial chromosome. As the DNA molecules present in natural yeast chromosomes are several hundred kilobases in length, it might be possible with an artificial chromosome to clone large pieces of DNA.

The structure and use of a YAC vector

Several YAC vectors have been developed but each one is constructed along the same lines, with pYAC3 being a typical example (Figure 7.8(a)). At first glance pYAC3 does not look much like an artificial chromosome, but on closer examination its unique features become apparent. pYAC3 is essentially a pBR322 plasmid into which a number of yeast genes have been inserted. Two of these genes, *URA3* and *TRP1*, have been encountered already as the selectable markers for YIp5 and YRp7 respectively. As in YRp7, the DNA fragment that carries *TRP1* also contains an origin of replication, but in pYAC3 this fragment is extended even further to include the

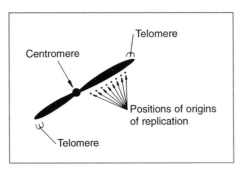

Figure 7.7 Chromosome structure.

Figure 7.8 A YAC vector and the way it is used to clone large pieces of DNA.

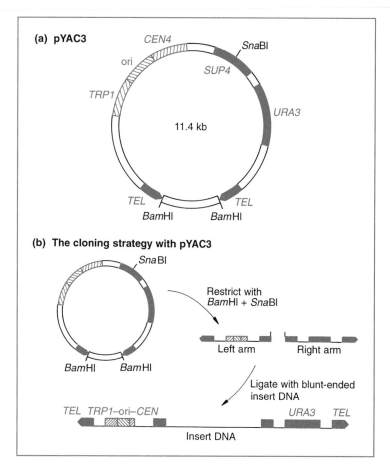

sequence called *CEN4*, which is the DNA from the centromere region of chromosome 4. The *TRP1*–origin–*CEN4* fragment therefore contains two of the three components of the artificial chromosome.

The third component, the telomeres, is provided by the two sequences called *TEL*. These are not themselves complete telomere sequences, but once inside the yeast nucleus they act as seeding sequences onto which telomeres will be built. This just leaves one other part of pYAC3 that has not been mentioned: *SUP4*, which is the selectable marker into which new DNA is inserted during the cloning experiment.

The cloning strategy with pYAC3 is as follows (Figure 7.8(b)). The vector is first restricted with a combination of *Bam*HI and *Sna*BI, cutting the molecule into three fragments. The *Bam*HI fragment is removed, leaving two arms, each bounded by one *TEL* sequence and one *Sna*BI site. The DNA to be cloned, which must have blunt ends (*Sna*BI is a blunt end cutter, recognizing the sequence TACGTA), is ligated between the two arms, producing the artificial chromosome. Protoplast transformation (p. 101) is then used to introduce the artificial chromosome into *S. cerevisiae*. The yeast strain that is used is a double auxotrophic mutant, $trp1^-$ $ura3^-$, which is converted to $trp1^+$ $ura3^+$ by the two markers on the artificial chromosome. Transformants are therefore selected by plating onto minimal medium, on which only cells containing a correctly constructed artificial chromosome are able to grow. Any cell transformed with an incorrect artificial chromosome, containing two left or two right arms rather than one of each, is not able to grow on minimal medium as one of the markers is absent. The presence of the insert DNA in the vector can be checked by testing for insertional inactivation of *SUP4*, which is carried out by a simple colour test: white colonies are recombinants, red colonies are not.

Applications for YAC vectors

The initial stimulus in designing artificial chromosomes came from yeast geneticists who wanted to use them to study various aspects of chromosome structure and behaviour, for instance to examine the segregation of chromosomes during meiosis. These experiments established that artificial chromosomes are stable during propagation in yeast cells and raised the possibility that they might be used as vehicles for genes that are too long to be cloned as a single fragment in an *E. coli* vector. Several important mammalian genes are greater than 100 kb in length (e.g. the human cystic fibrosis gene is 250 kb), beyond the capacity of all but the most sophisticated *E. coli* cloning systems (p. 127), but well within the range of a YAC vector. Yeast artificial chromosomes therefore opened the way to studies of the functions and modes of expression of genes that had previously been intractable to analysis by recombinant DNA techniques. A new dimension to these experiments has recently been provided by the discovery that under some circumstances YACs can be propagated in mammalian cells, enabling the functional analysis to be carried out in the organism in which the gene normally resides.

Yeast artificial chromosomes are equally important in the production of gene libraries. Recall that with fragments of 300 kb, the maximum insert size for the highest capacity *E. coli* vector, some 30 000 clones are needed for a human gene library (p. 128). However, YAC vectors are routinely used to clone 600 kb fragments, and special types are able to handle DNA up to 1400 kb in length, the latter bringing the size of a human gene library down to just 6500 clones. Unfortunately these 'mega-YACs' have run into problems with insert stability, the cloned DNA sometimes becoming rearranged by intramolecular recombination. Nevertheless, YACs have been of immense value in providing long pieces of cloned DNA for use in large scale DNA sequencing projects.

7.1.6 Vectors for other yeasts and fungi

Cloning vectors for other species of yeast and fungi are needed for basic studies of the molecular biology of these organisms and to extend the possible uses of yeasts and fungi in biotechnology. Episomal plasmids based on the *S. cerevisiae* 2 μm plasmid are able to replicate in a few other types of yeast, but the range of species is not broad enough for 2 μm vectors to be of general value. In any case, the requirements of biotechnology are better served by integrative plasmids, equivalent to YIps, as these provide stable recombinants that can be grown for long periods in bioreactors (p. 271). Efficient integrative vectors are now available for a number of species, including yeasts such as *Pichia pastoris* and *Kluveromyces lactis*, and the filamentous fungi *Aspergillus nidulans* and *Neurospora crassa*.

7.2 Cloning vectors for higher plants

Cloning vectors for higher plants were developed in the 1980s and their use has led to the **genetically modified (GM) crops** that are in the headlines today. We will examine the genetic modification of crops and other plants in Chapter 15. Here we look at the cloning vectors and how they are used.

Three types of vector system have been used with varying degrees of success with higher plants:

(1) Vectors based on naturally occurring plasmids of *Agrobacterium*.
(2) Direct gene transfer using various types of plasmid DNA.
(3) Vectors based on plant viruses.

7.2.1 *Agrobacterium tumefaciens* – nature's smallest genetic engineer

Although no naturally occurring plasmids are known in higher plants, one bacterial plasmid, the Ti plasmid of *Agrobacterium tumefaciens*, is of great importance.

A. tumefaciens is a soil microorganism that causes crown gall disease in many species of dicotyledonous plants. Crown gall occurs when a wound on the stem allows *A. tumefaciens* bacteria to invade the plant. After infection the bacteria cause a cancerous proliferation of the stem tissue in the region of the crown (Figure 7.9).

The ability to cause crown gall disease is associated with the presence of the Ti (**t**umour **i**nducing) plasmid within the bacterial cell. This is a large (greater than 200 kb) plasmid that carries numerous genes involved in the infective process (Figure 7.10(a)). A remarkable feature of the Ti plasmid is that, after infection, part of the molecule is integrated into the plant chromosomal DNA (Figure 7.10(b)). This segment, called the **T-DNA**, is between 15 and 30 kb in size, depending on the strain. It is maintained in a stable form in the plant cell and is passed on to daughter cells as an integral part of the chromosomes. But the most remarkable feature of the Ti plasmid is that the T-DNA contains eight or so genes that are expressed in the plant cell and are responsible for the cancerous properties of the transformed cells. These

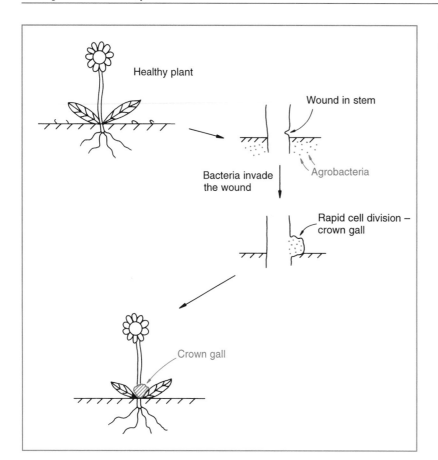

Figure 7.9 Crown gall disease.

genes also direct synthesis of unusual compounds, called opines, that the bacteria use as nutrients (Figure 7.10(c)). In short, *A. tumefaciens* genetically engineers the plant cell for its own purposes.

Using the Ti plasmid to introduce new genes into a plant cell

It was realized very quickly that the Ti plasmid could be used to transport new genes into plant cells. All that would be necessary would be to insert the new genes into the T-DNA and then the bacterium could do the hard work of integrating them into the plant chromosomal DNA. In practice this has proved quite a tricky proposition, mainly because the large size of the Ti plasmid makes manipulation of the molecule very difficult.

The main problem is of course that a unique restriction site is an impossibility with a plasmid 200 kb in size. Novel strategies have to be developed for inserting new DNA into the plasmid. Two are in general use:

(1) **The binary vector strategy** (Figure 7.11) is based on the observation that the T-DNA does not need to be physically attached to the rest of the Ti plasmid. A two plasmid system, with the T-DNA on a relatively small

Figure 7.10 The Ti plasmid and its integration into the plant chromosomal DNA after *A. tumefaciens* infection.

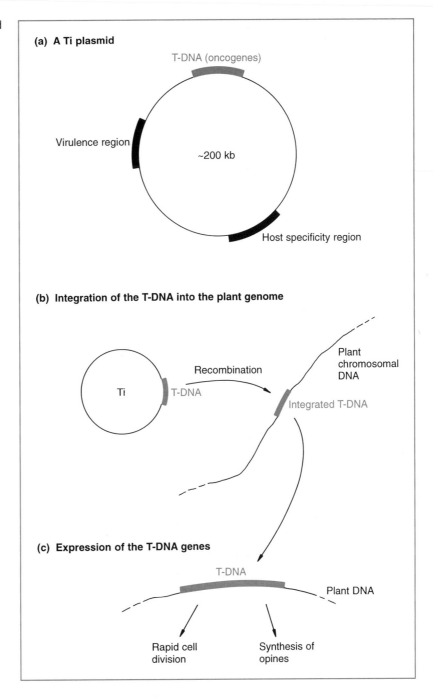

(a) A Ti plasmid

T-DNA (oncogenes)

Virulence region

~200 kb

Host specificity region

(b) Integration of the T-DNA into the plant genome

Ti

T-DNA

Recombination

Plant chromosomal DNA

Integrated T-DNA

(c) Expression of the T-DNA genes

T-DNA

Plant DNA

Rapid cell division

Synthesis of opines

molecule, and the rest of the plasmid in normal form, is just as effective at transforming plant cells. In fact some strains of *A. tumefaciens*, and related Agrobacteria, have natural binary plasmid systems. The T-DNA plasmid is small enough to have a unique restriction site and to be manipulated using standard techniques.

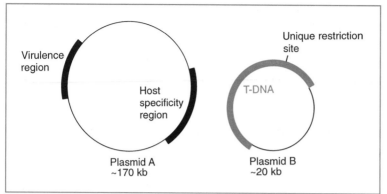

Figure 7.11 The binary vector strategy. Plasmids A and B complement each other when present together in the same *A. tumefaciens* cell. The T-DNA carried by plasmid B is transferred to the plant chromosomal DNA by proteins coded by genes carried by plasmid A.

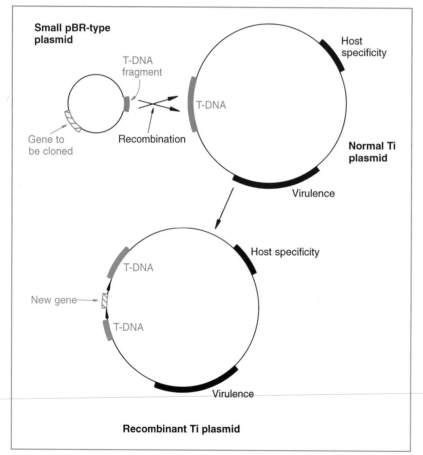

Figure 7.12 The cointegration strategy.

(2) **The cointegration strategy** (Figure 7.12) uses an entirely new plasmid, based on pBR322 or a similar *E. coli* vector, but carrying a small portion of the T-DNA. The homology between the new molecule and the Ti plasmid means that if both are present in the same *A. tumefaciens* cell, recombination can integrate the pBR plasmid into the T-DNA region.

The gene to be cloned is therefore inserted into a unique restriction site on the small pBR plasmid, introduced into *A. tumefaciens* cells carrying a Ti plasmid, and the natural recombination process left to integrate the new gene into the T-DNA. Infection of the plant leads to insertion of the new gene, along with the rest of the T-DNA, into the plant chromosomes.

Production of transformed plants with the Ti plasmid

If *A. tumefaciens* bacteria that contain an engineered Ti plasmid are introduced into a plant in the natural way, by infection of a wound in the stem, then only the cells in the resulting crown gall will possess the cloned gene (Figure 7.13(a)). This is obviously of little value to the biotechnologist. Instead a way of introducing the new gene into every cell in the plant is needed.

There are several solutions, the simplest being to infect not the mature plant but a culture of plant cells or protoplasts (p. 101) in liquid medium (Figure 7.13(b)). Plant cells and protoplasts whose cell walls have re-formed can be treated in the same way as microorganisms: for example, they can be plated onto a selective medium in order to isolate transformants. A mature plant regenerated from transformed cells will contain the cloned gene in every cell and will pass the cloned gene to its offspring.

However, regeneration of a transformed plant can occur only if the Ti vector has been '**disarmed**' so that the transformed cells do not display cancerous properties. Disarming is possible because the cancer genes, all of which lie in the T-DNA, are not needed for the infection process; infectivity is controlled mainly by the virulence region of the Ti plasmid. In fact the only parts of the T-DNA that are involved in infection are two 25 bp repeat sequences found at the left and right borders of the region integrated into the plant DNA. Any DNA placed between these two repeat sequences will be treated as 'T-DNA' and transferred to the plant. It is therefore possible to remove all the cancer genes from the normal T-DNA, and replace them with an entirely new set of genes, without disturbing the infection process.

A number of disarmed Ti cloning vectors are now available, a typical example being the binary vector pBIN19 (Figure 7.14). The left and right T-DNA borders present in this vector flank a copy of the *lacZ'* gene, containing a number of cloning sites, and a kanamycin resistance gene that functions after integration of the vector sequences into the plant chromosome. As with a yeast shuttle vector (p. 131), the initial manipulations that result in insertion of the gene to be cloned into pBIN19 are carried out in *E. coli*, the correct recombinant pBIN19 molecule then being transferred to *A. tumefaciens* and thence into the plant. Transformed plant cells are selected by plating onto agar medium containing kanamycin.

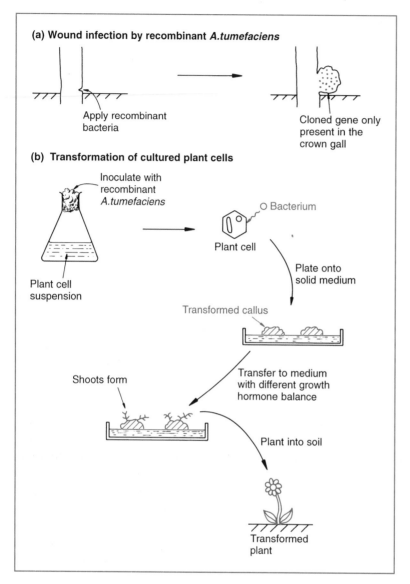

(a) Wound infection by recombinant *A.tumefaciens*

Apply recombinant bacteria

Cloned gene only present in the crown gall

(b) Transformation of cultured plant cells

Inoculate with recombinant *A.tumefaciens*

O Bacterium

Plant cell

Plant cell suspension

Plate onto solid medium

Transformed callus

Shoots form

Transfer to medium with different growth hormone balance

Plant into soil

Transformed plant

Figure 7.13 Transformation of plant cells by recombinant *A. tumefaciens*. (a) Infection of a wound: transformed plant cells are present only in the crown gall. (b) Transformation of a cell suspension: all the cells in the resulting plant are transformed.

The Ri plasmid

Over the years there has also been interest in developing plant cloning vectors based on the **Ri plasmid** of *Agrobacterium rhizogenes*. Ri and Ti plasmids are very similar, the main difference being that transfer of the T-DNA from an Ri plasmid to a plant results not in a crown gall but in hairy root disease, typified by a massive proliferation of a highly branched root system. The possibility of growing transformed roots at high density in liquid culture has been explored by biotechnologists as a potential means of obtaining large amounts of protein from genes cloned in plants (p. 290).

Figure 7.14 The binary Ti vector pBIN19. *kan^R* = kanamycin resistance gene.

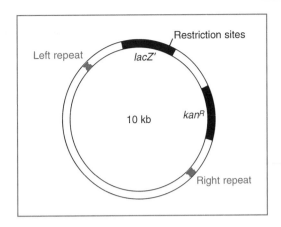

Limitations of cloning with *Agrobacterium* plasmids

Higher plants are divided into two broad categories, the monocots and the dicots. Several factors have combined to make it much easier to clone genes in dicots such as tomato, tobacco, potato, peas and beans, but much more difficult to obtain the same results with monocots. This has been frustrating because monocots include wheat, barley, rice and maize, which are the most important crop plants and hence the most desirable targets for genetic engineering projects.

The main difficulty stems from the fact that in nature *A. tumefaciens* and *A. rhizogenes* infect only dicotyledonous plants; monocots are outside of the normal host range. For some time it was thought that this natural barrier was insurmountable and that monocots were totally resistant to transformation with Ti and Ri vectors, but eventually artificial techniques for achieving T-DNA transfer were devised. But this was not the end of the story. Transformation with an *Agrobacterium* vector normally involves regeneration of an intact plant from a transformed protoplast, cell or callus culture. The ease with which a plant can be regenerated depends very much on the particular species involved and, once again, the most difficult plants are the monocots. Attempts to circumvent this problem have centred on the use of biolistics – bombardment with microprojectiles (p. 103) – to introduce plasmid DNA directly into plant embryos. Although this is a fairly violent transformation procedure it does not appear to be too damaging for the embryos, which still continue their normal development programme to produce mature plants. The approach has been successful with maize and several other important monocots.

7.2.2   Cloning genes in plants by direct gene transfer

Biolistics circumvents the need to use *Agrobacterium* as the means of transferring DNA into the plant cells. **Direct gene transfer** takes the process one step further and dispenses with the Ti plasmid altogether.

Direct gene transfer is based on the observation, first made in 1984, that a supercoiled bacterial plasmid, although unable to replicate in a plant cell on its own, can become integrated by recombination into one of the plant chromosomes. The recombination event is poorly understood but is almost certainly distinct from the processes responsible for T-DNA integration. It is also distinct from the chromosomal integration of a yeast vector (p. 133), as there is no requirement for a region of homology between the bacterial plasmid and the plant DNA. In fact integration appears to occur randomly at any position in any of the plant chromosomes (Figure 7.15).

Direct gene transfer therefore makes use of supercoiled plasmid DNA, possibly a simple bacterial plasmid such as pBR322, into which an appropriate selectable marker (e.g. a kanamycin resistance gene) and the gene to be cloned have been inserted. Biolistics is frequently used to introduce the plasmid DNA into plant embryos, but if the species being engineered can be regenerated from protoplasts or single cells then other strategies, possibly more efficient than biolistics, are possible.

One method involves resuspending protoplasts in a viscous solution of polyethylene glycol, a polymeric, negatively charged compound that is

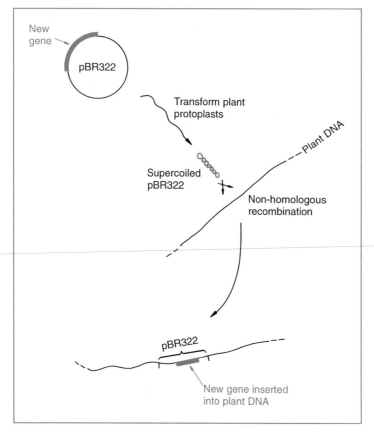

Figure 7.15 Direct gene transfer.

thought to precipitate DNA onto the surfaces of the protoplasts and to induce uptake by endocytosis (Figure 7.16(a)). Protoplasts can also be fused with DNA-containing liposomes (Figure 7.16(b)) or intact cells can be vigorously shaken with DNA-coated silica needles, which penetrate the cell wall and transfer the DNA into the interior.

After treatment, protoplasts are left for a few days in a solution that encourages regeneration of the cell walls. The cells are then spread onto selective medium to identify transformants and to provide callus cultures

Figure 7.16 Direct gene transfer by (a) precipitation of DNA onto the surfaces of protoplasts, and (b) fusion of protoplasts with DNA-containing liposomes.

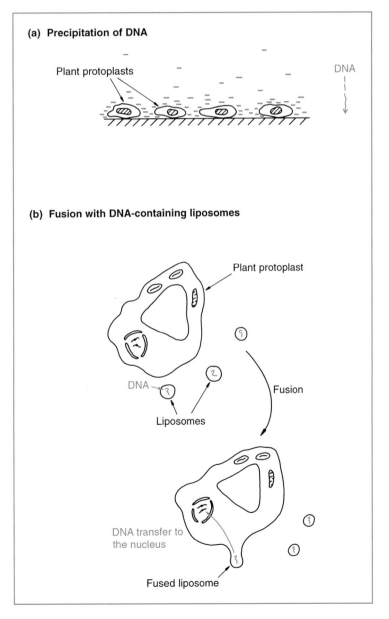

(a) Precipitation of DNA

Plant protoplasts

DNA

(b) Fusion with DNA-containing liposomes

Plant protoplast

DNA

Liposomes

Fusion

DNA transfer to the nucleus

Fused liposome

from which intact plants can be grown (exactly as described for the *Agrobacterium* system, Figure 7.13(b)).

7.2.3   Attempts to use plant viruses as cloning vectors

Modified versions of λ and M13 bacteriophages are important cloning vectors for *E. coli* (Chapter 6). Most plants are subject to viral infection, so could viruses be used to clone genes in plants? If they could they would be much more convenient to use than other types of vector, because with many viruses transformation can be achieved simply by rubbing the virus DNA onto the surface of a leaf. The natural infection process then spreads the virus throughout the plant.

The potential of plant viruses as cloning vectors has been explored for several years but, until recently, without great success. One problem is that the vast majority of plant viruses have genomes not of DNA but of RNA. RNA viruses are not so useful as potential cloning vectors because manipulations with RNA are rather more difficult to carry out. Only two classes of DNA virus are known to infect higher plants, the **caulimoviruses** and **geminiviruses**, and neither is ideally suited for gene cloning.

Caulimovirus vectors

Although one of the first successful plant genetic engineering experiments, back in 1984, used a caulimovirus vector to clone a new gene into turnip plants, two general difficulties with these viruses have limited their usefulness.

The first is that the total size of a caulimovirus genome is, like λ, constrained by the need to package it into its protein coat. Even after deletion of non-essential sections of the virus genome the capacity for carrying inserted DNA is still very limited. Recent research has shown that it might be possible to circumvent this problem by adopting a helper virus strategy, similar to that used with phagemids (p. 116). In this strategy, the cloning vector is a **cauliflower mosaic virus (CaMV)** genome that lacks several of the essential genes, which means that it can carry a large inserted gene but cannot by itself direct infection. Plants are inoculated with the vector DNA along with a normal CaMV genome. The normal viral genome provides the genes needed for the cloning vector to be packaged into virus proteins and spread through the plant.

This approach has considerable potential, but does not solve the second problem, which is the extremely narrow host range of caulimoviruses. This restricts cloning experiments to just a few plants, mainly brassicas such as turnips, cabbages and cauliflowers. Caulimoviruses have, however, been important in genetic engineering as the source of highly active promoters that work in all plants and that are used to obtain expression of genes introduced by Ti plasmid cloning or direct gene transfer.

Geminivirus vectors

What of the geminiviruses? These are particularly interesting because their natural hosts include plants such as maize and wheat and they could therefore be potential vectors for these and other monocots. But geminiviruses have presented their own set of difficulties, one problem being that during the infection cycle the genomes of some geminiviruses undergo rearrangements and deletions, which would scramble up any additional DNA that has been inserted, an obvious disadvantage for a cloning vector. Research over the years has addressed these problems and geminiviruses are starting to find specialist applications in plant cloning, with an increasingly important role anticipated in the future.

7.3   Cloning vectors for animals

Considerable effort has been put into the development of vector systems for cloning genes in animal cells. These vectors are needed in biotechnology for the synthesis of **recombinant protein** from genes that are not expressed correctly when cloned in *E. coli* or yeast (Chapter 13), and methods for cloning in humans are being sought by clinical molecular biologists attempting to devise techniques for **gene therapy** (p. 309), in which a disease is treated by introduction of a cloned gene into the patient.

The clinical aspect has meant that most attention has been directed at cloning systems for mammals, but important progress has also been made with insects. Cloning in insects is interesting because it makes use of a novel type of vector that we have not met so far. We will therefore examine insect vectors before concluding the chapter with an overview of the cloning methods used with mammals.

7.3.1   Cloning vectors for insects

The fruit fly, *Drosophila melanogaster*, has been and still is one of the most important model organisms used by biologists. Its potential was first recognized by the famous geneticist Thomas Hunt Morgan, who in 1910 started to carry out genetic crosses between fruit flies with different eye colours and other inherited characteristics. These experiments led to the techniques still used today for gene mapping in insects and other animals.

More recently, the discovery that the homeotic selector genes of *Drosophila* – the genes that control the overall body plan of the fly – are closely related to equivalent genes in mammals, has led to *D. melanogaster* being used as a model for the study of human developmental processes. The importance of the fruitfly in modern biology makes it imperative that vectors for cloning genes in this organism are available.

P elements as cloning vectors for *Drosophila*

The development of cloning vectors for *Drosophila* has taken a different route to that followed with bacteria, yeast, plants and mammals. No plasmids are known in *Drosophila* and although fruit flies are, like all organisms, susceptible to infection with viruses, these have not been used as the basis for cloning vectors. Instead, cloning in *Drosophila* makes use of a **transposon** called the **P element**.

Transposons are common in all types of organisms. They are short pieces of DNA (usually less than 10 kb in length) that can move from one position to another in the chromosomes of a cell. P elements, which are one of several types of transposon in *Drosophila*, are 2.9 kb in length and contain three genes flanked by short inverted repeat sequences at either end of the element (Figure 7.17(a)). The genes code for transposase, the enzyme that carries out the transposition process, and the inverted repeats form the recognition sequences that enable the enzyme to identify the two ends of the inserted transposon.

As well as moving from one site to another within a single chromosome, P elements can also jump between chromosomes, or between a plasmid carrying a P element and one of the fly's chromosomes (Figure 7.17(b)). The latter is the key to the use of P elements as cloning vectors. The vector is a plasmid that carries two P elements, one of which contains the insertion site for the DNA that will be cloned. Insertion of the new DNA into this P element results in disruption of its transposase gene, so this element is inactive. The second P element carried by the plasmid is therefore one that has an intact version of the transposase gene. Ideally this second element should not itself be transferred to the *Drosophila* chromosomes, so it has its 'wings clipped': its inverted repeats are removed so that the transposase does not recognize it as being a real P element. Once the gene to be cloned has been inserted into the vector, the plasmid DNA is microinjected into fruit fly embryos. The transposase from the wings-clipped P element directs transfer of the engineered P element into one of the fruit fly chromosomes. If this happens within a germline nucleus then the adult fly that develops from the embryo will carry copies of the cloned gene in all its cells. P element cloning was first developed in the 1980s and has made a number of important contributions to *Drosophila* genetics.

Cloning vectors based on insect viruses

Although virus vectors have not been developed for cloning genes in *Drosophila*, one type of virus, the **baculovirus**, has played an important role in gene cloning with other insects. The main use of baculovirus vectors is in the production of recombinant protein, and we will return to them when we consider this topic in Chapter 13.

Figure 7.17 Cloning in *Drosophila* with a P element vector. (a) The structure of a P element. (b) Transposition of a P element from a plasmid to a fly chromosome. (c) The structure of a P element cloning vector. The left-hand P element contains a cloning site (R) that disrupts its transposase gene. The right-hand P element has an intact transposase gene but cannot itself transpose because it is 'wings-clipped' – it lacks terminal inverted repeats.

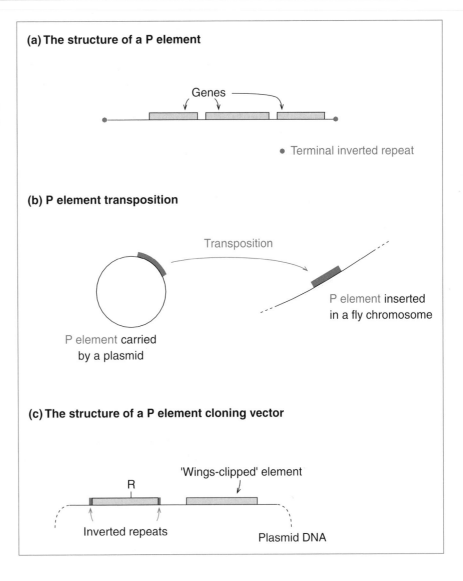

(a) The structure of a P element

Genes

• Terminal inverted repeat

(b) P element transposition

Transposition

P element carried by a plasmid

P element inserted in a fly chromosome

(c) The structure of a P element cloning vector

R

'Wings-clipped' element

Inverted repeats

Plasmid DNA

7.3.2

Cloning in mammals

At present, gene cloning in mammals is carried out for one of three reasons.

(1) To achieve a **gene knockout**, which is an important technique used to help determine the function of an unidentified gene (p. 262). These experiments are usually carried out with rodents such as mice.

(2) For production of recombinant protein in a mammalian cell culture, and in the related technique of **pharming**, which involves genetic engineering of a farm animal so that it synthesizes an important protein such as a pharmaceutical, often in its milk (p. 289).

(3) In **gene therapy**, in which human cells are engineered in order to treat a disease (p. 311).

Cloning vectors for mammals

For many years it was thought that viruses would prove to be the key to cloning in mammals. This expectation has only partially been realized. The first cloning experiment involving mammalian cells was carried out in 1979 with a vector based on simian virus 40 (SV40). This virus is capable of infecting several mammalian species, following a lytic cycle in some hosts and a lysogenic cycle in others. The genome is 5.2 kb in size (Figure 7.18(a)) and contains two sets of genes, the 'early' genes, expressed early in the infection cycle and coding for proteins involved in viral DNA replication, and the 'late' genes, coding for viral capsid proteins. SV40 suffers from the same problem as λ and the plant caulimoviruses, in that packaging constraints limit the amount of new DNA that can be inserted into the genome. Cloning with SV40 therefore involves replacing one or more of the existing genes with the DNA to be cloned. In the original experiment a segment of the late gene region was replaced (Figure 7.18(b)), but early gene replacement is also an option.

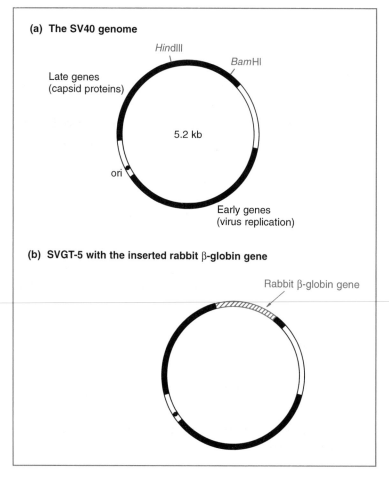

(a) **The SV40 genome**

HindIII

BamHI

Late genes
(capsid proteins)

5.2 kb

ori

Early genes
(virus replication)

(b) **SVGT-5 with the inserted rabbit β-globin gene**

Rabbit β-globin gene

Figure 7.18 SV40 and an example of its use as a cloning vector. To clone the rabbit β-globin gene the HindIII to BamHI restriction fragment was deleted (resulting in SVGT-5) and replaced with the rabbit gene.

In the years since 1979 a number of other types of virus have been used to clone genes in mammals. **Adenoviruses** enable larger fragments of DNA to be cloned than is possible with an SV40 vector, though they are more difficult to handle because the genomes are bigger. **Papillomaviruses**, which also have a relatively high capacity for inserted DNA, have the important advantage of enabling a stable transformed cell line to be obtained.

Many mammalian viruses kill their host cells soon after infection, so special tricks are needed if these are to be used for anything other than short-term transformation experiments. Bovine papillomavirus (BPV), which causes warts on cattle, is particularly attractive because it has an unusual infection cycle in mouse cells, taking the form of a multicopy plasmid with about 100 molecules present per cell. It does not cause the death of the mouse cell, and BPV molecules are passed to daughter cells on cell division. Shuttle vectors consisting of BPV and pBR322 sequences, and capable of replication in both mouse and bacterial cells, have been used for the production of recombinant proteins in mouse cell lines.

At present, **retroviruses** are the most commonly used vectors for cloning genes in mammalian cells. Their most important application is in gene therapy, and we will return to them when we discuss this topic (p. 311).

Gene cloning without a vector

One of the reasons why virus vectors have not become widespread in mammalian gene cloning is because it was discovered in the early 1990s that the most effective way of transferring new genes into mammalian cells is by microinjection. Although a difficult procedure to carry out, microinjection of bacterial plasmids, or linear DNA copies of genes, into mammalian nuclei results in the DNA being inserted into the chromosomes, possibly as multiple copies in a tandem, head-to-tail arrangement (Figure 7.19). This procedure is generally looked on as more satisfactory than the use of a viral vector because it avoids the possibility that viral DNA will infect the cells and cause defects of one kind or another.

A knockout mouse (p. 263), which has copies of the cloned gene in all its cells, can be generated by microinjection of a fertilized egg cell which is

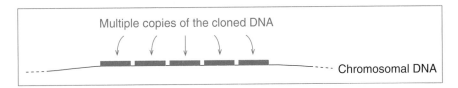

Figure 7.19 Multiple copies of cloned DNA molecules inserted as a tandem array in a chromosomal DNA molecule.

subsequently cultured *in vitro* for several cell divisions and then implanted into a foster mother. Alternatively an **embryonic stem (ES) cell** can be used. These are obtained from within an early embryo and, unlike most mammalian cells, are totipotent, meaning that their developmental pattern is not pre-set and cells descended from them can form many different structures in the adult mouse. After microinjection, the ES cell is placed back in an embryo which is implanted into the foster mother. The resulting mouse is a **chimera**, comprising a mixture of engineered and non-engineered cells, because the embryo that receives the ES cell also contains a number of ordinary cells that contribute, along with the ES cell, to the make-up of the adult mouse. Non-chimeric mice, which contain the cloned gene in all their cells, are obtained by allowing the chimera to reproduce, as some of the offspring will be derived from egg cells that contain the cloned gene.

Further reading

Anderson, C. (1993) Genome shortcut leads to problems. *Science*, **259**, 1684–7. Some of the problems with using YACs to clone large pieces of DNA.

Bevan, M. (1984) Binary *Agrobacterium* vectors for plant transformation. *Nucleic Acids Research*, **12**, 8711–21.

Brisson, N., Paszkowski, J., Penswick, J.R., Gromenborn, B., Potrykus, I. & Hohn, T. (1984) Expression of a bacterial gene in plants by using a viral vector. *Nature*, **310**, 511–14. [The first cloning experiment with a caulimovirus.]

Broach, J.R. (1982) The yeast 2 μm circle. *Cell*, **28**, 203–4.

Burke, D.T., Carle, G.F. & Olson, M.V. (1987) Cloning of large segments of exogenous DNA into yeast by means of artificial chromosome vectors. *Science*, **236**, 806–12.

Chilton, M.D. (1983) A vector for introducing new genes into plants. *Scientific American*, **248** (June), 50–59. The Ti plasmid.

Evans, M.J., Carlton, M.B.L. & Russ, A.P. (1997) Gene trapping and functional genomics. *Trends in Genetics*, **13**, 370–74. [Includes a description of the use of ES cells.]

Graham, F.L. (1990) Adenoviruses as expression vectors and recombinant vaccines. *Trends in Biotechnology*, **8**, 20–25.

Hamer, D.H. & Leder, P. (1979) Expression of the chromosomal mouse β-maj-globin gene cloned in SV40. *Nature*, **281**, 35–40.

Hansen, G. & Wright, M.S. (1999) Recent advances in the transformation of plants. *Trends in Plant Science*, **4**, 226–31.

Monaco, A.P. & Larin, Z. (1994) YACs, BACs, PACs and MACs: artificial chromosomes as research tools. *Trends in Biotechnology*, **12**, 280–86.

Nadolska-Orczyk, A., Orczyk, W. & Przetakiewicz, A. (2000) *Agrobacterium*-mediated transformation of cereals – from technique development to its application. *Acta Physiologiae Plantarum*, **22**, 77–78.

Parent, S.A. *et al.* (1985) Vector systems for the expression, analysis and cloning of DNA sequences in *S. cerevisiae*. *Yeast*, **1**, 83–138. [Details of different yeast cloning vectors.]

Paszkowski, J., Shillito, R.D., Saul, M. *et al.* (1984) Direct gene transfer to plants. *EMBO Journal*, **3**, 2717–22.

Rubin, G.M. & Spradling, A.C. (1982) Genetic transformation of *Drosophila* with transposable element vectors. *Science*, **218**, 348–353. [Cloning with P elements.]

Timmermans, M.C.P., Das, O.P. & Messing, J. (1994) Geminiviruses and their uses as extrachromosomal replicons. *Annual Review of Plant Physiology and Plant Molecular Biology*, **45**, 79–112.

Viaplana, R., Turner, D.S. & Covey, S.N. (2001) Transient expression of a GUS reporter gene from cauliflower mosaic virus replacement vectors in the presence and absence of helper virus. *Journal of General Virology*, **82**, 59–65. [The helper virus approach to cloning with CaMV.]

Chapter 8

How to Obtain a Clone of a Specific Gene

In the preceding chapters we have examined the basic methodology used to clone genes, and surveyed the range of vector types that are used with bacteria, yeast, plants and animals. Now we must look at the methods available for obtaining a clone of an individual, specified gene. This is the critical test of a gene cloning experiment, success or failure often depending on whether or not a strategy can be devised by which clones of the desired gene can be selected directly, or alternatively, distinguished from other recombinants. Once this problem has been resolved, and a clone has been obtained, the molecular biologist is able to make use of a wide variety of different techniques that will extract information about the gene. The most important of these will be described in Chapters 10 and 11.

8.1 The problem of selection

The problem faced by the molecular biologist wishing to obtain a clone of a single, specified gene was illustrated in Figure 1.4. Even the simplest organisms, such as *E. coli*, contain several thousand genes, and a restriction digest of total cell DNA produces not only the fragment carrying the desired gene, but also many other fragments carrying all the other genes (Figure 8.1 (a)). During the ligation reaction there is no selection for an individual fragment: numerous different recombinant DNA molecules are produced, all containing different pieces of DNA (Figure 8.1(b)). Consequently a variety of recombinant clones are obtained after transformation and plating out (Figure 8.1(c)). Somehow the correct one must be identified.

8.1.1 There are two basic strategies for obtaining the clone you want

Although there are many different procedures by which the desired clone can be obtained, all are variations on two basic themes.

Figure 8.1 The problem of selection.

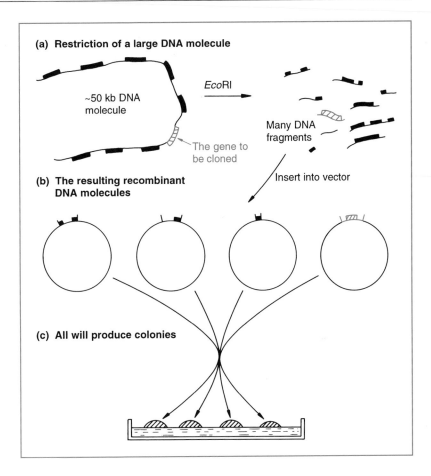

(a) Restriction of a large DNA molecule

~50 kb DNA molecule

*Eco*RI

The gene to be cloned

Many DNA fragments

Insert into vector

(b) The resulting recombinant DNA molecules

(c) All will produce colonies

(1) **Direct selection for the desired gene** (Figure 8.2(a)), which means that the cloning experiment is designed in such a way that the only clones that are obtained are clones of the required gene. Almost invariably, selection occurs at the plating out stage.

(2) **Identification of the clone from a gene library** (Figure 8.2(b)), which entails an initial 'shotgun' cloning experiment, to produce a clone library representing all or most of the genes present in the cell, followed by analysis of the individual clones to identify the correct one.

In general terms, direct selection is the preferred method, as it is quick and usually unambiguous. However, as we shall see, it is not applicable to all genes. Techniques for clone identification are therefore very important, especially as complete genomic libraries of many organisms are now available.

8.2   Direct selection

To be able to select for a cloned gene it is necessary to plate the transformants onto an agar medium on which only the desired recombinants, and no

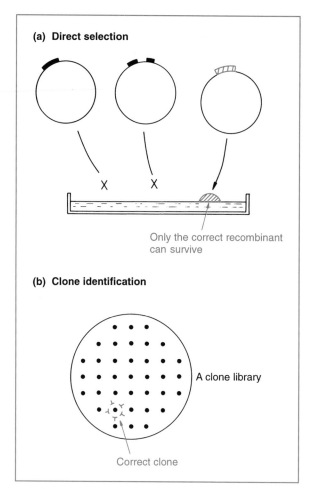

Figure 8.2 The basic strategies that can be used to obtain a particular clone. (a) Direct selection. (b) Identification of the desired recombinant from a clone library.

others, can grow. The only colonies that are obtained will therefore be ones that comprise cells containing the desired recombinant DNA molecule.

The simplest example of direct selection occurs when the desired gene specifies resistance to an antibiotic. As an example we will consider an experiment to clone the gene for kanamycin resistance from plasmid R6-5. This plasmid carries genes for resistances to four antibiotics: kanamycin, chloramphenicol, streptomycin and sulphonamide. The kanamycin resistance gene lies within one of the 13 *Eco*RI fragments (Figure 8.3(a)).

To clone this gene the *Eco*RI fragments of R6-5 would be inserted into the *Eco*RI site of a vector such as pBR322. The ligated mix will comprise many copies of 13 different recombinant DNA molecules, one set of which carries the gene for kanamycin resistance (Figure 8.3(b)).

Insertional inactivation cannot be used to select recombinants when the *Eco*RI site of pBR322 is used. This is because this site does not lie in either the ampicillin or the tetracycline resistance genes of this plasmid (Figure 6.1). But this is immaterial for cloning the kanamycin resistance gene because in

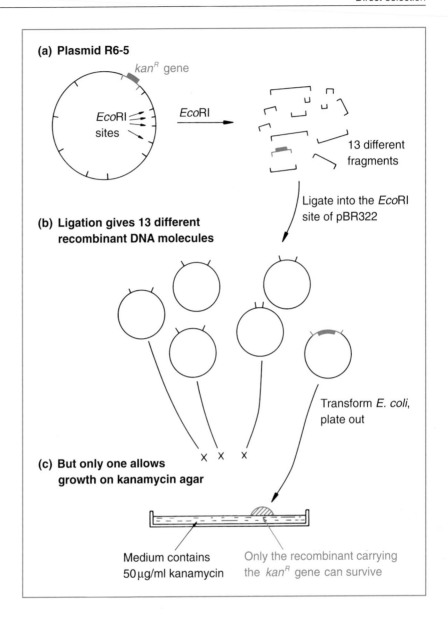

Figure 8.3 Direct selection for the cloned R6-5 kanamycin resistance (*kan^R*) gene.

(a) Plasmid R6-5

kan^R gene

EcoRI sites

EcoRI

13 different fragments

Ligate into the *Eco*RI site of pBR322

(b) Ligation gives 13 different recombinant DNA molecules

Transform *E. coli*, plate out

(c) But only one allows growth on kanamycin agar

X X X

Medium contains 50 μg/ml kanamycin

Only the recombinant carrying the *kan^R* gene can survive

this case the cloned gene can be used as the selectable marker. Transformants are plated onto kanamycin agar, on which the only cells able to survive and produce colonies are those recombinants that contain the cloned kanamycin resistance gene (Figure 8.3(c)).

8.2.1 Marker rescue extends the scope of direct selection

Direct selection would be very limited indeed if it could be used only for cloning antibiotic resistance genes. Fortunately the technique can be extended by making use of mutant strains of *E. coli* as the hosts for transformation.

As an example, consider an experiment to clone the gene *trpA* from *E. coli*. This gene codes for the enzyme tryptophan synthase, which is involved in biosynthesis of the essential amino acid tryptophan. A mutant strain of *E. coli* that has a non-functional *trpA* gene is called *trpA⁻*, and is able to survive only if tryptophan is added to the growth medium. *E. coli trpA⁻* is therefore another example of an auxotroph (p. 131).

This *E. coli* mutant can be used to clone the correct version of the *trpA* gene. Total DNA is first purified from a normal (wild-type) strain of the bacterium. Digestion with a restriction endonuclease, followed by ligation into a vector, produces numerous recombinant DNA molecules, one of which may, with luck, carry an intact copy of the *trpA* gene (Figure 8.4(a)). This is of course the functional gene as it has been obtained from the wild-type strain.

The ligation mixture is now used to transform the auxotrophic *E. coli trpA⁻* cells (Figure 8.4(b)). The vast majority of the resulting transformants will be auxotrophic, but a few now have the plasmid-borne copy of the correct *trpA* gene. These recombinants are non-auxotrophic – they no longer

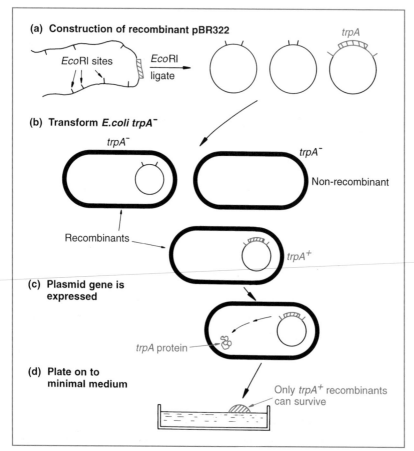

(a) Construction of recombinant pBR322

EcoRI sites

EcoRI
ligate

trpA

(b) Transform E.coli trpA⁻

trpA⁻

trpA⁻

Non-recombinant

Recombinants

trpA⁺

(c) Plasmid gene is expressed

trpA protein

(d) Plate on to minimal medium

Only *trpA⁺* recombinants can survive

Figure 8.4 Direct selection for the *trpA* gene cloned in a *trpA⁻* strain of *E. coli*.

require tryptophan as the cloned gene is able to direct production of tryptophan synthase (Figure 8.4(c)). Direct selection is therefore performed by plating transformants onto minimal medium, which lacks any added supplements, and in particular has no tryptophan (Figure 8.4(d)). Auxotrophs cannot grow on minimal medium, so the only colonies to appear are recombinants that contain the cloned *trpA* gene.

8.2.2 The scope and limitations of marker rescue

Although marker rescue can be used to obtain clones of many genes, the technique is subject to two limitations.

(1) A mutant strain must be available for the gene in question.
(2) A medium on which only the wild-type can survive is needed.

Marker rescue is applicable for most genes that code for biosynthetic enzymes, as clones of these genes can be selected on minimal medium in the manner described for *trpA*. However, the technique is not limited to *E. coli* nor even bacteria; auxotrophic strains of yeast and filamentous fungi are also available, and marker rescue has been used to select genes cloned into these organisms.

In addition, *E. coli* auxotrophs can be used as hosts for the selection of some genes from other organisms. Often there is sufficient similarity between equivalent enzymes from different bacteria, or even from yeast, for the foreign enzyme to function in *E. coli*, so that the cloned gene is able to transform the host to wild-type.

8.3 Identification of a clone from a gene library

Although marker rescue is a powerful technique it is not all-embracing and there are many important genes that cannot be selected by this method. Many bacterial mutants are not auxotrophs, so the mutant and wild-type strains cannot be distinguished by plating onto minimal or any other special medium. In addition, neither marker rescue nor any other direct selection method is of much use in providing bacterial clones of genes from higher organisms (i.e. animals and plants), as in these cases the differences are usually so great that the foreign enzymes do not function in the bacterial cell.

The alternative strategy must therefore be considered. This is where a large number of different clones are obtained and the desired one identified in some way.

8.3.1 Gene libraries

Before looking at the methods used to identify individual clones, the library itself must be considered. A genomic library (p. 126) is a collection of clones

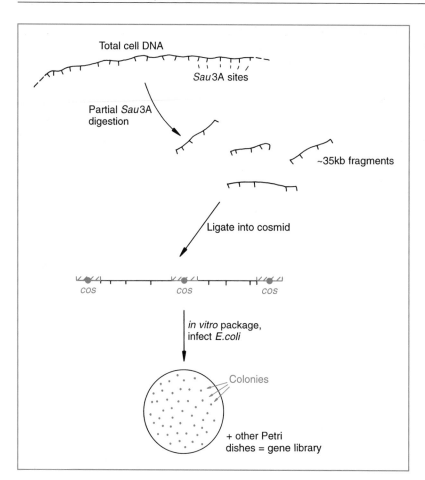

Figure 8.5 Preparation of a gene library in a cosmid vector.

sufficient in number to be likely to contain every single gene present in a particular organism. Genomic libraries are prepared by purifying total cell DNA, and then making a partial restriction digest, resulting in fragments that can be cloned into a suitable vector (Figure 8.5), usually a λ replacement vector, a cosmid, or possibly a yeast artificial chromosome (YAC), bacterial artificial chromosome (BAC) or P1 vector.

For bacteria, yeast and fungi, the number of clones needed for a complete genomic library is not so large as to be unmanageable (Table 6.1). For plants and animals, though, a complete library contains so many different clones that identification of the desired one may prove a mammoth task. With these organisms a second type of library, specific not to the whole organism but to a particular cell type, may be more useful.

8.3.2 Not all genes are expressed at the same time

A characteristic of most multicellular organisms is specialization of individual cells. A human being, for example, is made up of a large number of different cell types – brain cells, blood cells, liver cells, etc. Each cell contains the same complement of genes, but in different cell types different sets of genes are switched on, while others are silent (Figure 8.6).

The fact that only relatively few genes are expressed in any one type of cell can be utilized in preparation of a library if the material that is cloned is not DNA but messenger RNA (mRNA). Only those genes that are being expressed are transcribed into mRNA, so if mRNA is used as the starting material then the resulting clones comprise only a selection of the total number of genes in the cell.

A cloning method that uses mRNA would be particularly useful if the desired gene is expressed at a high rate in an individual cell type. For example, the gene for gliadin, one of the nutritionally important proteins present in wheat, is expressed at a very high level in the cells of developing wheat seeds. In these cells over 30% of the total mRNA specifies gliadin. Clearly, if we could clone the mRNA from wheat seeds we would obtain a large number of clones specific for gliadin.

Figure 8.6 Different genes are expressed in different types of cell.

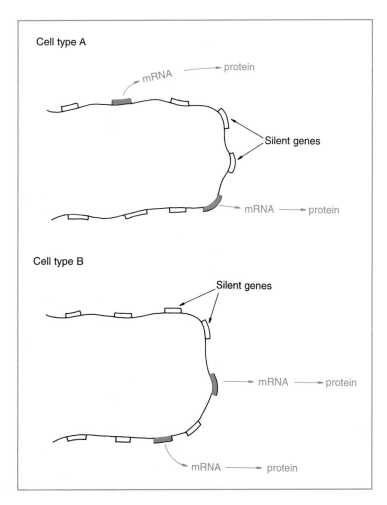

8.3.3

mRNA can be cloned as complementary DNA

Messenger RNA cannot itself be ligated into a cloning vector. However, mRNA can be converted into DNA by **complementary DNA (cDNA)** synthesis.

The key to this method is the enzyme reverse transcriptase (p. 58) which synthesizes a DNA polynucleotide strand complementary to an existing RNA strand (Figure 8.7(a)). Once the cDNA strand has been synthesized the RNA member of the hybrid molecule can be partially degraded by treating with ribonuclease (RNase) H (Figure 8.7(b)). The remaining RNA fragments then serve as primers (p. 57) for DNA polymerase I, which synthesizes the second cDNA strand (Figure 8.7(c)), resulting in a double-stranded DNA fragment that can be ligated into a vector and cloned (Figure 8.7(d)).

The resulting cDNA clones are representative of the mRNA present in the original preparation. In the case of mRNA prepared from wheat seeds, the cDNA library would contain a large proportion of clones representing gliadin mRNA (Figure 8.7(e)). Other clones will also be present, but locating the cloned gliadin cDNA is a much easier process than identifying the equivalent gene from a complete wheat genomic library.

8.4

Methods for clone identification

Once a suitable library has been prepared, a number of procedures can be employed to attempt identification of the desired clone. Although a few of these procedures are based on detection of the translation product of the cloned gene, it is usually easier to identify directly the correct recombinant DNA molecule. This can be achieved by the important technique of **hybridization probing**.

8.4.1

Complementary nucleic acid strands hybridize to each other

Any two single-stranded nucleic acid molecules have the potential to form base pairs with one another. With most pairs of molecules the resulting hybrid structures are unstable, as only a small number of individual interstrand bonds are formed (Figure 8.8(a)). However, if the polynucleotides are complementary, extensive base pairing can occur to form a stable double-stranded molecule (Figure 8.8(b)). Not only can this occur between single-stranded DNA molecules to form the DNA double helix, but also between single-stranded RNA molecules and between combinations of one DNA and one RNA strand (Figure 8.8(c)).

Nucleic acid hybridization can be used to identify a particular recombinant clone if a DNA or RNA probe, complementary to the desired gene, is available. The exact nature of the probe will be discussed later in the chapter. First we must consider the technique itself.

Figure 8.7 One possible scheme for cDNA cloning (see text for details). Poly(A) = polyadenosine, oligo(dT) = oligodeoxythymidine.

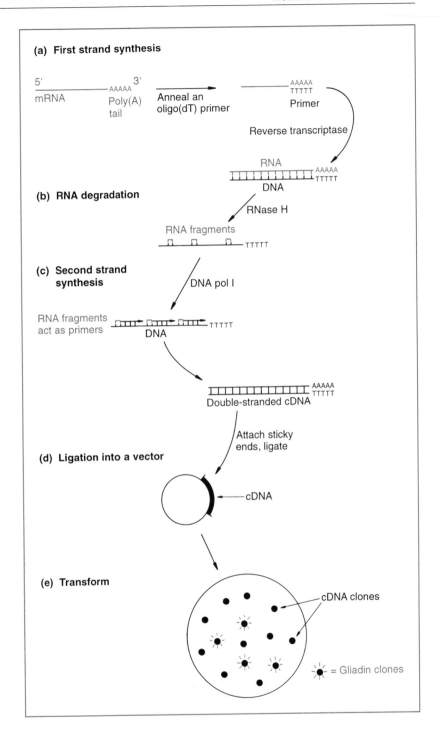

(a) **First strand synthesis**

(b) **RNA degradation**

(c) **Second strand synthesis**

(d) **Ligation into a vector**

(e) **Transform**

Colony and plaque hybridization probing

Hybridization probing can be used to identify recombinant DNA molecules contained in either bacterial colonies or bacteriophage plaques. Thanks to innovative techniques developed in the late 1970s, it is not necessary

of the sequence of the cloned gene. If the gene itself is not available (which presumably is the case if the aim of the experiment is to provide a clone of it), then what can be used as the probe?

In practice the nature of the probe is determined by the information available about the desired gene. We will consider three possibilities.

(1) Where the desired gene is expressed at a high level in a cell type from which a cDNA clone library has been prepared.

(2) Where the amino acid sequence of the protein coded by the gene is completely or partially known.

(3) Where the gene is a member of a family of related genes.

Abundancy probing to analyse a cDNA library

As described earlier in this chapter, a cDNA library is often prepared in order to obtain a clone of a gene expressed at a relatively high level in a particular cell type. In the example of a cDNA library from developing wheat seeds, a large proportion of the clones are copies of the mRNA transcripts of the gliadin gene (Figure 8.7(e)).

Identification of the gliadin clones is simply a case of using individual cDNA clones from the library to probe all the other members of the library (Figure 8.12). A clone is selected at random and the recombinant DNA

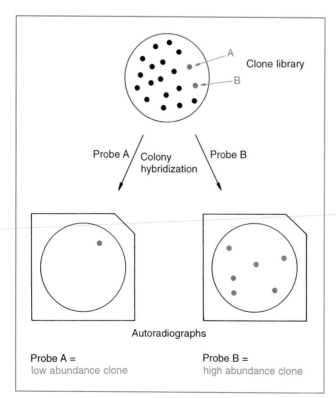

Figure 8.12 Probing within a library to identify an abundant clone.

molecule purified, labelled and used to probe the remaining clones. This is repeated with different clones as probes until one that hybridizes to a large proportion of the library is obtained. This abundant cDNA is considered a possible gliadin clone and analysed in greater detail (e.g. by DNA sequencing and isolation of the translation product) to confirm the identification.

Oligonucleotide probes for genes whose translation products have been characterized

Often the gene to be cloned codes for a protein that has already been studied in some detail. In particular the amino acid sequence of the protein might have been determined, using sequencing techniques that have been available for over 40 years. If the amino acid sequence is known, then it is possible to use the genetic code to predict the nucleotide sequence of the relevant gene. This prediction is always an approximation, as only methionine and tryptophan can be assigned unambiguously to triplet codons; all other amino acids are coded by at least two codons each. Nevertheless, in most cases, the different codons for an individual amino acid are related. Alanine, for example, is coded by GCA, GCT, GCG and GCC, so two out of the three nucleotides of the triplet coding for alanine can be predicted with certainty.

As an example to clarify how these predictions are made, consider cytochrome c, a protein that plays an important role in the respiratory chain of all aerobic organisms. The cytochrome c protein from yeast was sequenced in 1963, with the result shown in Figure 8.13. This sequence contains a segment, starting at amino acid 59, that runs Trp–Asp–Glu–Asn–Asn–Met. The genetic code states that this hexapeptide is coded by TGG–GA$_C^T$–GA$_G^A$–AA$_C^T$–AA$_C^T$–ATG. Although this represents a total of 16 different possible sequences, 14 of the 18 nucleotides can be predicted with certainty.

Oligonucleotides of up to about 50 nucleotides in length can easily be synthesized in the laboratory (Figure 8.14). An oligonucleotide probe could therefore be constructed according to the predicted nucleotide sequence, and this probe might be able to identify the gene coding for the protein in question. In the example of yeast cytochrome c, the 16 possible oligonucleotides

Figure 8.13 The amino acid sequence of yeast cytochrome c. The hexapeptide that is highlighted red is the one used to illustrate how a nucleotide sequence can be predicted from an amino acid sequence.

```
                                                              15
GLY–SER–ALA–LYS–LYS–GLY–ALA–THR–LEU–PHE–LYS–THR–ARG–CYS–GLU–
                                                              30
LEU–CYS–HIS–THR–VAL–GLU–LYS–GLY–GLY–PRO–HIS–LYS–VAL–GLY–PRO–
                                                              45
ASN–LEU–HIS–GLY–ILE–PHE–GLY–ARG–HIS–SER–GLY–GLN–ALA–GLN–GLY–
                                                              60
TYR–SER–TYR–THR–ASP–ALA–ASN–ILE–LYS–LYS–ASN–VAL–LEU–TRP–ASP–
                                                              75
GLU–ASN–ASN–MET–SER–GLU–TYR–LEU–THR–ASN–PRO–LYS–LYS–TYR–ILE–
                                                              90
PRO–GLY–THR–LYS–MET–ALA–PHE–GLY–GLY–LEU–LYS–LYS–GLU–LYS–ASP–
                                                             103
ARG–ASN–ASP–LEU–ILE–THR–TYR–LEU–LYS–LYS–ALA–CYS–GLU
```

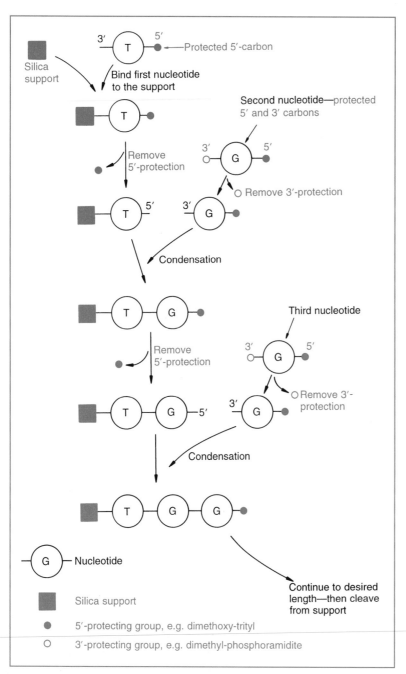

Figure 8.14 A simplified scheme for oligonucleotide synthesis. The protecting groups attached to the 3′ and 5′ termini prevent reactions between individual mononucleotides. By carefully controlling the time at which the protecting groups are removed, mononucleotides can be added one by one to the growing oligonucleotide.

that can code for Trp–Asp–Glu–Asn–Asn–Met would be synthesized, either separately or as a pool, and then used to probe a yeast genomic or cDNA library (Figure 8.15). One of the oligonucleotides in the probe will have the correct sequence for this region of the cytochrome c gene, and its hybridization signal will indicate which clones carry this gene.

Figure 8.15 The use of a synthetic, end-labelled oligonucleotide to identify a clone of the yeast cytochrome c gene.

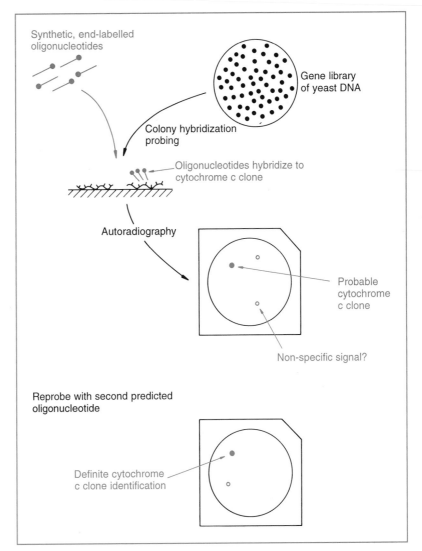

The result can be checked by carrying out a second probing with a mixture of oligonucleotides whose sequences are predicted from a different segment of the cytochrome c protein (Figure 8.15). However, the segment of the protein used for nucleotide sequence prediction must be chosen with care: the hexapeptide Ser–Glu–Tyr–Leu–Thr–Asn, which immediately follows our first choice, could be coded by several thousand different 18-nucleotide sequences, clearly an unsuitable choice for a synthetic probe.

Heterologous probing allows related genes to be identified

Often a substantial amount of nucleotide similarity is seen when two genes for the same protein, but from different organisms, are compared, a reflection of the conservation of gene structure during evolution. Frequently, two genes from related organisms are sufficiently similar for a single-stranded probe prepared from one gene to form a stable hybrid with the second gene. Although the two molecules are not entirely complementary, enough base pairs are formed to produce a stable structure (Figure 8.16(a)).

Heterologous probing makes use of hybridization between related sequences for clone identification. For example, the yeast cytochrome c gene, identified in the previous section by oligonucleotide probing, could itself be used as a hybridization probe to identify cytochrome c genes in clone libraries of other organisms. A probe prepared from the yeast gene would not be entirely complementary to the gene from, say, *Neurospora crassa*, but sufficient base pairing should occur for a hybrid to be formed and be detected by autoradiography (Figure 8.16(b)). The experimental conditions would be modified so that the heterologous structure is not destabilized and lost before autoradiography.

Heterologous probing can also identify related genes in the *same* organism. If the wheat gliadin cDNA clone, identified earlier in the chapter by abundancy probing, is used to probe a genomic library, it will hybridize not

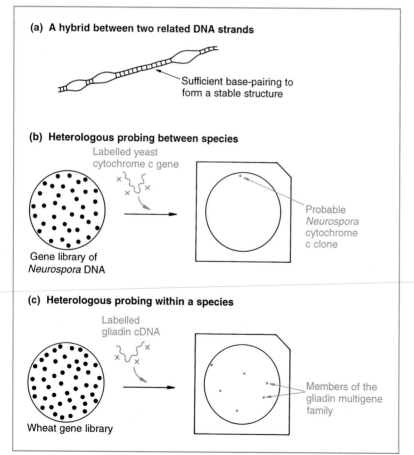

(a) A hybrid between two related DNA strands

Sufficient base-pairing to form a stable structure

(b) Heterologous probing between species

Labelled yeast cytochrome c gene

Gene library of *Neurospora* DNA

Probable *Neurospora* cytochrome c clone

(c) Heterologous probing within a species

Labelled gliadin cDNA

Wheat gene library

Members of the gliadin multigene family

Figure 8.16 Heterologous probing.

only to its own gene but to a variety of other genes as well (Figure 8.16(c)). These are all related to the gliadin cDNA, but have slightly different nucleotide sequences. This is because the wheat gliadins form a complex group of related proteins that are coded by the members of a **multigene family**. Once one gene in the family has been cloned, then all the other members can be isolated by heterologous probing.

8.4.4 Identification methods based on detection of the translation product of the cloned gene

Hybridization probing is usually the preferred method for identification of a particular recombinant from a clone library. The technique is easy to perform and, with modifications introduced in recent years, can be used to check up to 10 000 recombinants per experiment, allowing large genomic libraries to be screened in a reasonably short time. Nevertheless, the requirement for a probe that is at least partly complementary to the desired gene sometimes makes it impossible to use hybridization in clone identification. On these occasions a different strategy is needed.

The main alternative to hybridization probing is **immunological screening**. The distinction is that, whereas with hybridization probing the cloned DNA fragment is itself directly identified, an immunological method detects the protein coded by the cloned gene. Immunological techniques therefore presuppose that the cloned gene is being expressed, so that the

Figure 8.17 Antibodies. (a) Antibodies in the bloodstream bind to foreign molecules and help degrade them. (b) Purified antibodies can be obtained from a small volume of blood taken from a rabbit injected with the foreign protein.

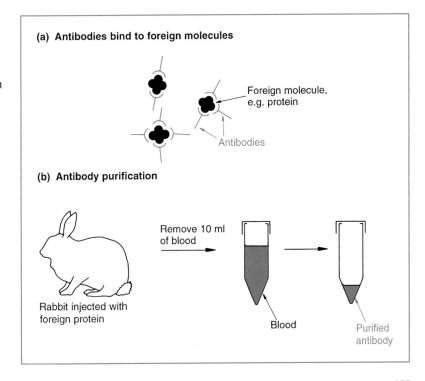

(a) Antibodies bind to foreign molecules

Foreign molecule, e.g. protein

Antibodies

(b) Antibody purification

Rabbit injected with foreign protein

Remove 10 ml of blood

Blood

Purified antibody

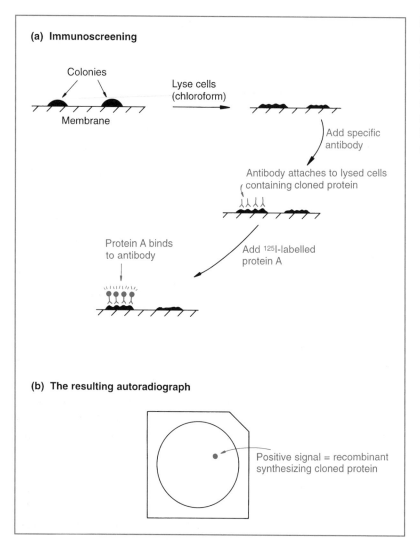

(a) **Immunoscreening**

Colonies

Membrane

Lyse cells
(chloroform)

Add specific
antibody

Antibody attaches to lysed cells
containing cloned protein

Protein A binds
to antibody

Add ^{125}I-labelled
protein A

(b) **The resulting autoradiograph**

Positive signal = recombinant
synthesizing cloned protein

Figure 8.18 Using a purified antibody to detect protein in recombinant colonies. Instead of labelled protein A, the antibody itself can be labelled, or alternatively a second labelled antibody which binds specifically to the primary antibody can be used.

protein is being made, and that this protein is not normally present in the host cells.

Antibodies are required for immunological detection methods

If a purified sample of a protein is injected into the bloodstream of a rabbit, the immune system of the animal responds by synthesizing antibodies that bind to and help degrade the foreign molecule (Figure 8.17(a)). This is a version of the natural defence mechanism that the animal uses to deal with invasion by bacteria, viruses and other infective agents.

Once a rabbit is challenged with a protein, the levels of antibody present in its bloodstream remain high enough over the next few days for substantial quantities to be purified. It is not necessary to kill the rabbit, because as little as 10 ml of blood provides a considerable amount of antibody (Figure 8.17(b)). This purified antibody binds only to the protein with which the animal was originally challenged.

Using a purified antibody to detect protein in recombinant colonies

There are several versions of immunological screening, but the most useful method is a direct counterpart of colony hybridization probing. Recombinant colonies are transferred to a polyvinyl membrane, the cells are lysed, and a solution containing the specific antibody is added (Figure 8.18(a)). Either the antibody itself is labelled, or the membrane is subsequently washed with a solution of labelled **protein A**, a bacterial protein that specifically binds to the immunoglobulins that antibodies are made of (Figure 8.18(b)). The label can be a radioactive one, in which case the colonies that bind the label are detected by autoradiography, or non-radioactive labels resulting in a fluorescent or chemiluminescent signal can be used.

The problem of gene expression

Immunological screening depends on the cloned gene being expressed so that the protein translation product is present in the recombinant cells. However, as will be discussed in greater detail in Chapter 13, a gene from one organism is often not expressed in a different organism. In particular, it is very unlikely that a cloned animal or plant gene (with the exception of chloroplast genes) will be expressed in *E. coli* cells. This problem can be circumvented by using a special type of vector, called an **expression vector** (p. 273), designed specifically to promote expression of the cloned gene in a bacterial host. Immunological screening of recombinant *E. coli* colonies carrying animal genes cloned into expression vectors has in fact been very useful in obtaining genes for several important hormones.

Further reading

Benton, W.D. & Davis, R.W. (1977) Screening λgt recombinant clones by hybridization to single plaques in situ. *Science*, **196**, 180–82.

Dyson, N.J. & Brown, T.A. (2001) Immobilization of nucleic acids and hybridization analysis. In: *Essential Molecular Biology: A Practical Approach*, Vol. 2 (ed. T.A. Brown), 2nd edn. *In press.* IRL Press at Oxford University Press, Oxford.

Feinberg, A.P. & Vogelstein, B. (1983) A technique for labelling DNA restriction fragments to high specific activity. *Analytical Biochemistry*, **132**, 6–13. [Random priming labelling.]

Grunstein, M. & Hogness, D.S. (1975) Colony hybridization: a method for the isolation of cloned cDNAs that contain a specific gene. *Proceedings of the National Academy of Sciences of the USA*, **72**, 3961–5.

Gubler, U. & Hoffman, B.J. (1983) A simple and very efficient method for generating cDNA libraries. *Gene*, **25**, 263–9.

Thorpe, G.H.G., Kricka, L.J., Moseley, S.B. & Whitehead, T.P. (1985) Phenols as enhancers of the chemiluminescent horseradish peroxidase–luminol–hydrogen peroxide reaction: application in luminescence-monitored enzyme immunoassays. *Clinical Chemistry*, **31**, 1335–41. [Describes the basis to a non-radioactive labelling method.]

Young, R.A. & Davis, R.W. (1983) Efficient isolation of genes by using antibody probes. *Proceedings of the National Academy of Sciences of the USA*, **80**, 1194–8.

Chapter 9

The Polymerase Chain Reaction

As a result of the last seven chapters we have become familiar not only with the basic principles of gene cloning, but also with fundamental molecular biology techniques such as restriction analysis, gel electrophoresis, DNA labelling and DNA–DNA hybridization. To complete our basic education in DNA analysis we must now return to the second major technique for studying genes, the polymerase chain reaction (PCR). PCR is a very uncomplicated technique: all that happens is that a short region of a DNA molecule, a single gene for instance, is copied many times by a DNA polymerase enzyme (Figure 1.2). This might seem a rather trivial exercise, but it has a multitude of applications in genetics research and in broader areas of biology.

We begin this chapter with an outline of the polymerase chain reaction in order to understand exactly what it achieves. Then we will run through the relevant methodology, following the steps involved in PCR and the special methods that have been devised for studying the amplified DNA fragments that are obtained.

9.1 The polymerase chain reaction in outline

The polymerase chain reaction results in the selective amplification of a chosen region of a DNA molecule. Any region of any DNA molecule can be chosen, so long as the sequences at the borders of the region are known. The border sequences must be known because in order to carry out a PCR, two short oligonucleotides must hybridize to the DNA molecule, one to each strand of the double helix (Figure 9.1). These oligonucleotides, which act as primers for the DNA synthesis reactions, delimit the region that will be amplified.

Amplification is usually carried out by the DNA polymerase I enzyme from *Thermus aquaticus*. As mentioned on p. 57, this organism lives in hot

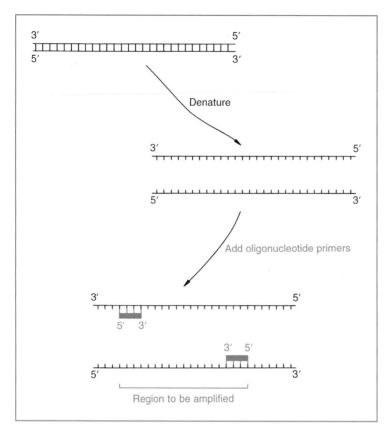

Figure 9.1 Hybridization of the oligonucleotide primers to the template DNA at the beginning of a PCR.

springs, and many of its enzymes, including *Taq* polymerase, are thermostable, meaning that they are resistant to denaturation by heat treatment. As will be apparent in a moment, the thermostability of *Taq* polymerase is an essential requirement in PCR methodology.

To begin a PCR amplification, the enzyme is added to the primed template DNA and incubated so that it synthesizes new complementary strands (Figure 9.2(a)). The mixture is then heated to 94°C so that the newly synthesized strands detach from the template (Figure 9.2(b)), and cooled, enabling more primers to hybridize at their respective positions, including positions on the newly synthesized strands. *Taq* polymerase, which unlike most types of DNA polymerase is not inactivated by the heat treatment, now carries out a second round of DNA synthesis (Figure 9.2(c)). The cycle of denaturation–hybridization–synthesis is repeated, usually 25–30 times, resulting in the eventual synthesis of several hundred million copies of the amplified DNA fragment (Figure 9.2(d)).

At the end of a PCR a sample of the reaction mixture is usually analysed by agarose gel electrophoresis, sufficient DNA having been produced for the amplified fragment to be visible as a discrete band after staining with

Figure 9.2 The polymerase chain reaction. dNTPs = 2'-deoxynucleotide 5'-triphosphates.

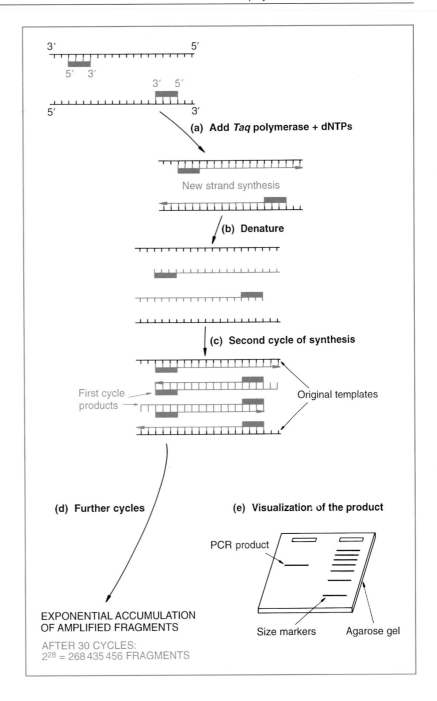

ethidium bromide (Figure 9.2(e)). This may by itself provide useful information about the DNA region that has been amplified, or alternatively the PCR product can be ligated into a plasmid or bacteriophage vector, cloned in the normal way, and examined by standard techniques such as DNA sequencing.

9.2

PCR in more detail

Although PCR experiments are very easy to set up, they must be planned carefully if the results are to be of any value. The sequences of the primers are critical to the success of the experiment, as are the precise temperatures used in the heating and cooling stages of the reaction cycle. Also there is the important question of what can be done with the amplified DNA molecules once they have been obtained.

9.2.1

Designing the oligonucleotide primers for a PCR

The primers are the key to the success or failure of a PCR experiment. If the primers are designed correctly the experiment results in amplification of a single DNA fragment, corresponding to the target region of the template molecule. If the primers are incorrectly designed the experiment will fail, possibly because no amplification occurs, or possibly because the wrong fragment, or more than one fragment, is amplified (Figure 9.3). Clearly a great deal of thought must be put into the design of the primers.

Working out appropriate sequences for the primers is not a problem: they must correspond with the sequences flanking the target region on the template molecule. Each primer must of course be complementary (not identical) to its template strand in order for hybridization to occur, and the 3′ ends of the hybridized primers should point towards one another (Figure 9.4). The DNA fragment to be amplified should not be greater than about 3 kb in length and ideally is less than 1 kb. Fragments up to 10 kb can be amplified by standard PCR techniques, but the longer the fragment the less efficient the amplification and the more difficult it is to obtain consistent results. Amplification of very long fragments – up to 40 kb – is possible but requires special methods.

The first important issue to address is the length of the primers. If the primers are too short they might hybridize to non-target sites and give

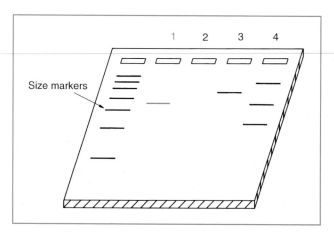

Figure 9.3 The results of PCRs with well designed and poorly designed primers. Lane 1 shows a single amplified fragment of the expected size, the result of a well designed experiment. In lane 2 there is no amplification product, suggesting that one or both of the primers were unable to hybridize to the template DNA. Lanes 3 and 4 show, respectively, an amplification product of the wrong size, and a mixture of products (the correct product plus two wrong ones); both results are due to hybridization of one or both of the primers to non-target sites on the template DNA molecule.

Size markers

Figure 9.4 A pair of primers designed to amplify the human α_1-globin gene. The exons of the gene are shown as closed boxes, the introns as open boxes.

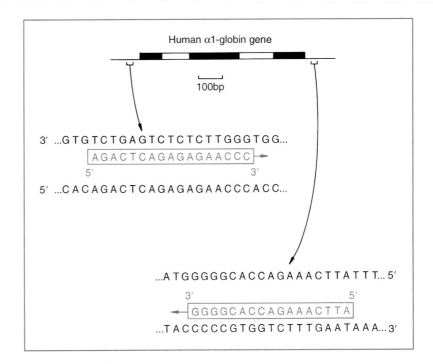

undesired amplification products. To illustrate this point, imagine that total human DNA is used in a PCR experiment with a pair of primers eight nucleotides in length (in PCR jargon these are called '8-mers'). The likely result is that a number of different fragments will be amplified. This is because attachment sites for these primers are expected to occur, on average, once every $4^8 = 65\,536$ bp, giving approximately 46 000 possible sites in the 3 000 000 kb of nucleotide sequence that makes up the human genome. This means that it would be very unlikely that a pair of 8-mer primers would give a single, specific amplification product with human DNA (Figure 9.5(a)).

What if the 17-mer primers shown in Figure 9.4 are used? The expected frequency of a 17-mer sequence is once every $4^{17} = 17\,179\,869\,184$ bp. This figure is over five times greater than the length of the human genome, so a 17-mer primer would be expected to have just one hybridization site in total human DNA. A pair of 17-mer primers should therefore give a single, specific amplification product (Figure 9.5(b)).

Why not simply make the primers as long as possible? Because the length of the primer influences the rate at which it hybridizes to the template DNA; long primers hybridize at a slower rate. The efficiency of the PCR, measured by the number of amplified molecules produced during the experiment, is therefore reduced if the primers are too long, as complete hybridization to the template molecules cannot occur in the time allowed during the reaction cycle. In practice, primers longer than 30-mer are rarely used.

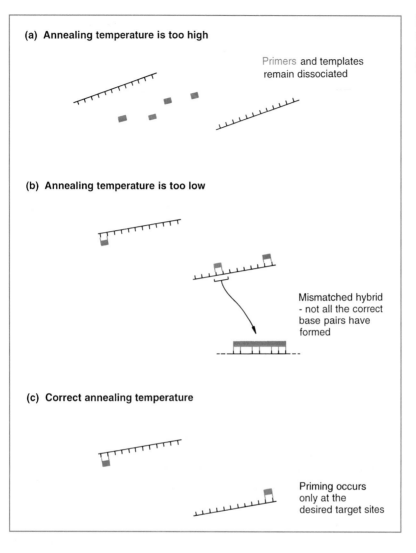

(a) **Annealing temperature is too high**

Primers and templates remain dissociated

(b) **Annealing temperature is too low**

Mismatched hybrid - not all the correct base pairs have formed

(c) **Correct annealing temperature**

Priming occurs only at the desired target sites

Figure 9.7 Temperature has an important effect on the hybridization of the primers to the template DNA.

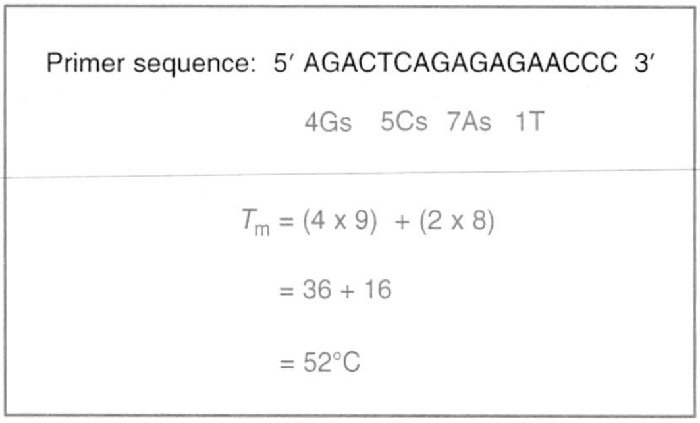

Figure 9.8 Calculating the T_m of a primer.

Primer sequence: 5' AGACTCAGAGAGAACCC 3'

4Gs   5Cs   7As   1T

$T_m = (4 \times 9) + (2 \times 8)$

$= 36 + 16$

$= 52°C$

so that they have identical T_ms. If this is not the case, the appropriate annealing temperature for one primer may be too high or too low for the other member of the pair.

9.2.3   After the PCR: studying PCR products

Polymerase chain reaction is often the starting point for a longer series of experiments in which the amplification product is studied in various ways in order to gain information about the DNA molecule that acted as the original template. We will encounter many studies of this type in Parts 2 and 3, when we examine the applications of gene cloning and DNA analysis in research and biotechnology. Although a wide range of procedures have been devised for studying PCR products, three techniques are particularly important.

(1)   Gel electrophoresis of PCR products.
(2)   Cloning of PCR products.
(3)   Sequencing of PCR products.

The first two of these techniques are dealt with in this chapter. The third technique is deferred until Chapter 10, when all aspects of DNA sequencing will be covered.

Gel electrophoresis of PCR products

As indicated in Figure 9.2, the results of most PCR experiments are checked by running a portion of the amplified reaction mixture in an agarose gel. A band representing the amplified DNA may be visible after ethidium bromide staining, or if the DNA yield is low the product can be detected by Southern hybridization (p. 198). If the expected band is absent, or if additional bands are present, something has gone wrong and the experiment must be repeated.

In some cases, agarose gel electrophoresis is used not only to determine if a PCR experiment has worked, but also to obtain additional information. For example, the presence of restriction sites in the amplified region of the template DNA can be determined by treating the PCR product with a restriction endonuclease before running the sample in the agarose gel (Figure 9.9). This is a type of **restriction fragment length polymorphism (RFLP) analysis** and is important both in the construction of genome maps (p. 255) and in studying genetic diseases (p. 306).

Alternatively, the exact size of the PCR product can be used to establish if the template DNA contains an insertion or deletion mutation in the amplified region (Figure 9.9). Length mutations of this type form the basis of **DNA profiling**, a central technique in forensic science (Chapter 16).

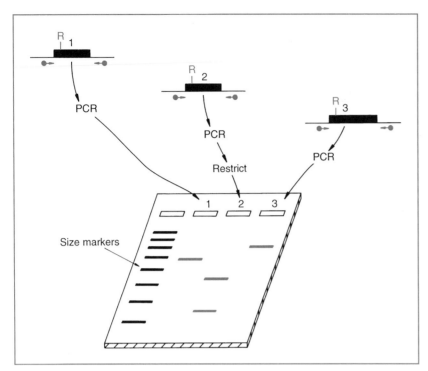

Figure 9.9 Gel electrophoresis of the PCR product can provide information on the template DNA molecule. Lanes 1 and 2 show, respectively, an unrestricted PCR product and a product restricted with the enzyme that cuts at site R. Lane 3 shows the result obtained when the template DNA contains an insertion in the amplified region.

In some experiments the mere presence or absence of the PCR product is the diagnostic feature: for example, when PCR is used as the screening procedure to identify a desired gene from a genomic or complementary DNA (cDNA) library. Carrying out PCRs with every clone in a genomic library might seem to be a tedious task, but one of the advantages of PCR is that individual experiments are quick to set up and many PCRs can be performed in parallel. The workload can also be reduced by **combinatorial screening**, an example of which is shown in Figure 9.10.

Cloning PCR products

Some applications require that after a PCR the resulting products are ligated into a vector and examined by any of the standard methods used for studying cloned DNA. This may sound easy but there are complications.

The first problem concerns the ends of the PCR products. From an examination of Figure 9.2 it might be imagined that fragments amplified by PCR are blunt-ended. If this was the case they could be inserted into a cloning vector by blunt end ligation, or alternatively the PCR products could be provided with sticky ends by the attachment of linkers or adaptors (p. 78). Unfortunately, the situation is not so straightforward. *Taq* polymerase tends to add an additional nucleotide, usually an adenosine, to the end of each strand that it synthesizes. This means that a double-stranded PCR product is not blunt-ended, and instead most 3′ termini have a single

Figure 9.10
Combinatorial screening of clones in microtitre trays. A library of 960 clones is screened by a series of PCRs, each with a combination of clones. The clone combinations that give positive results enable the well(s) containing positive clone(s) to be identified. For example, if positive PCRs are given with row A of tray 2, row D of tray 6, column 7 of tray 2, and column 9 of tray 6, then it can be deduced that there are positive clones in well A7 of tray 2 and well D9 of tray 6. This deduction is unambiguous and can be made without carrying out the 'well' PCRs (which would have given positive results for wells A7 and D9). The 'well' PCRs are only needed if there are two positive clones in the same tray. (Redrawn with permission from Brown, T.A. (1999) *Genomes*. BIOS Scientific Publishers, Oxford.)

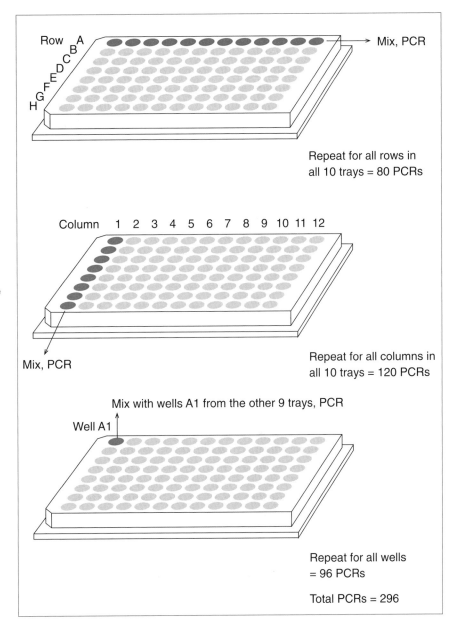

Repeat for all rows in all 10 trays = 80 PCRs

Repeat for all columns in all 10 trays = 120 PCRs

Repeat for all wells = 96 PCRs

Total PCRs = 296

Figure 9.11 Polynucleotides synthesized by *Taq* polymerase often have an extra adenosine at their 3′ ends.

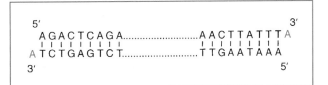

nucleotide overhang (Figure 9.11). The overhangs could be removed by treatment with an exonuclease enzyme, resulting in PCR products with true blunt ends, but this is not a popular approach as it is difficult to prevent the exonuclease from becoming overactive and causing further damage to the ends of the molecules.

One solution is to use a special cloning vector which carries thymidine (T) overhangs and which can therefore be ligated to a PCR product (Figure 9.12). These vectors are usually prepared by restricting a standard vector at a blunt end site, and then treating with *Taq* polymerase in the presence of just 2′-deoxythymidine 5′-triphosphate (dTTP). No primer is present so all the polymerase can do is add a T nucleotide to the 3′ ends of the blunt-ended vector molecule, resulting in the T-tailed vector into which the PCR products can be inserted.

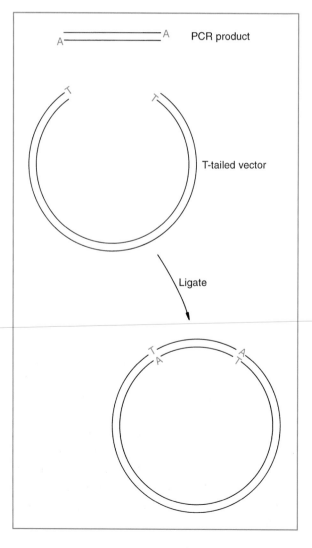

Figure 9.12 Using a special T-tailed vector to clone a PCR product.

Figure 9.13 Obtaining a PCR product with a sticky end through use of a primer whose sequence includes a restriction site.

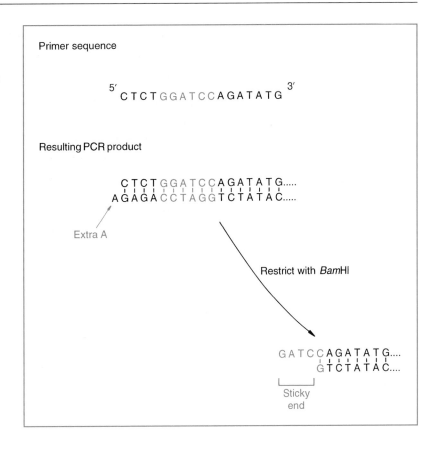

Primer sequence

5′ CTCTGGATCCAGATATG 3′

Resulting PCR product

CTCTGGATCCAGATATG.....
AGAGACCTAGGTCTATAC.....

Extra A

Restrict with *Bam*HI

GATCCAGATATG....
GTCTATAC....

Sticky end

Figure 9.14 A PCR primer with a restriction site present within an extension at the 5′ end.

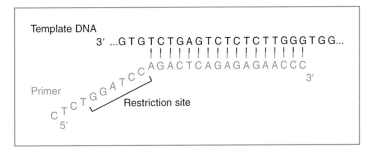

Template DNA

3′ ...GTGTCTGAGTCTCTCTTGGGTGG...

AGACTCAGAGAGAACCC 3′

Primer

ATCC
CTGGA
CTCT
5′

Restriction site

A second solution is to design primers that contain restriction sites. After PCR the products are treated with the restriction endonuclease, which cuts each molecule within the primer sequence, leaving sticky-ended fragments that can be ligated efficiently into a standard cloning vector (Figure 9.13). The approach is not limited to those instances where the primers span restriction sites that are present in the template DNA. Instead, the restriction site can be included within a short extension at the 5′ end of each primer (Figure 9.14). These extensions cannot hybridize to the template molecule, but they are copied during the PCR, resulting in PCR products that carry terminal restriction sites.

9.3

Problems with the error rate of *Taq* polymerase

All DNA polymerases make mistakes during DNA synthesis, occasionally inserting an incorrect nucleotide into the growing DNA strand. Most polymerases, however, are able to rectify these errors by reversing over the mistake and resynthesizing the correct sequence. This property is referred to as the 'proofreading' function and depends on the polymerase possessing a 3′ to 5′ exonuclease activity (p. 53).

Taq polymerase appears to lack a proofreading activity and as a result is unable to correct its errors. This means that the DNA synthesized by *Taq*

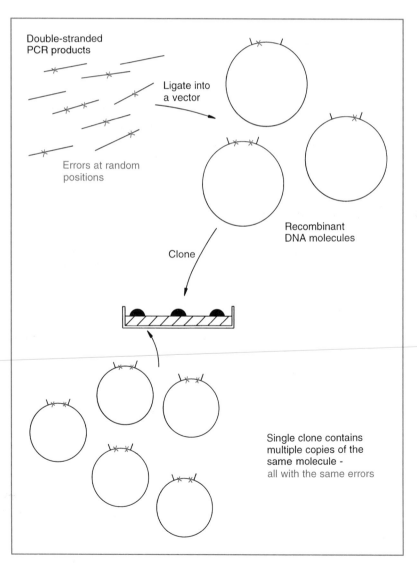

Figure 9.15 The high error rate of *Taq* polymerase becomes a factor when PCR products are cloned.

Double-stranded
PCR products

Ligate into
a vector

Errors at random
positions

Recombinant
DNA molecules

Clone

Single clone contains
multiple copies of the
same molecule -
all with the same errors

polymerase is not always an accurate copy of the template molecule. The error rate has been estimated at 1 mistake for every 9000 nucleotides of DNA that is synthesized, which might appear to be almost insignificant but which translates to 1 error in every 300 bp for the PCR products obtained after 30 cycles. This is because PCR involves copies being made of copies of copies, so the polymerase-induced errors gradually accumulate, the fragments produced at the end of a PCR containing copies of earlier errors along with any new errors introduced during the final round of synthesis.

For many applications this high error rate does not present a problem. In particular, direct sequencing of a PCR product (p. 210) provides the correct sequence of the template, even though the PCR products contain the errors introduced by *Taq* polymerase. This is because the errors are distributed randomly, so for every molecule that has an error at a particular nucleotide position, there will be many molecules with the correct sequence. In this context the error rate is indeed insignificant.

This is not the case if the PCR products are cloned. Each resulting clone contains multiple copies of a single amplified fragment, so the cloned DNA does not necessarily have the same sequence as the original template molecule used in the PCR (Figure 9.15). This possibility lends an uncertainty to all experiments carried out with cloned PCR products and dictates that, whenever possible, the amplified DNA should be studied directly rather than being cloned.

Further reading

Marchuk, D., Drumm, M., Saulino, A. & Collins, F.S. (1991) Construction of T-vectors, a rapid and general system for direct cloning of unmodified PCR products. *Nucleic Acids Research*, **19**, 1154.

Rychlik, W., Spencer, W.J. & Rhoads, R.E. (1990) Optimization of the annealing temperature for DNA amplification *in vitro. Nucleic Acids Research*, **18**, 6409–12.

Saiki, R.K., Gelfand, D.H., Stoffel, S. *et al.* (1988) Primer-directed enzymatic amplification of DNA with a thermostable DNA polymerase. *Science*, **239**, 487–91. [The first description of PCR with *Taq* polymerase.]

PART 2
THE APPLICATIONS OF GENE CLONING AND DNA ANALYSIS IN RESEARCH

Studying Gene Location and Structure

Part 1 of this book has shown how a skilfully performed cloning or polymerase chain reaction (PCR) experiment can provide a pure sample of an individual gene, or any other DNA sequence, separated from all the other genes and DNA sequences in the cell. Now we can turn our attention to the ways in which cloning, PCR and other DNA analysis techniques are used to study genes. We will consider three aspects of molecular biology research:

(1)　The techniques used to study the location and structure of a gene (this Chapter).

(2)　The methods used to study the expression and function of a gene (Chapter 11).

(3)　The various techniques that collectively are called **genomics** and **post-genomics** (Chapter 12).

10.1　How to study the location of a gene

Several techniques are available for determining the location of a gene within a DNA molecule. The exact nature of the procedure used depends on the size of the DNA molecule involved, with the techniques applicable for small molecules, such as normal and recombinant versions of plasmids and phage chromosomes, being different from those used for gene location on the large DNA molecules contained in eukaryotic chromosomes.

10.1.1　Locating the position of a gene on a small DNA molecule

Consider again the example of direct selection used in Chapter 8, which resulted in the kanamycin resistance gene from R6-5 being cloned as an *Eco*RI fragment carried by pBR322 (p. 158). Now the clone is available it would be useful to know within which of the 13 R6-5 *Eco*RI fragments the gene is located, as this information would allow the gene to be placed on the

R6-5 restriction map and to be positioned relative to other genes on this plasmid.

First, an *Eco*RI restriction digest of R6-5 must be electrophoresed in an agarose gel so that the individual fragments can be seen (Figure 10.1(a)). One of these fragments is the same as that inserted into the recombinant pBR322 molecule that carries the kanamycin resistance gene. The aim is therefore to label the recombinant molecule and use it to probe the restriction digest. This can be attempted while the restriction fragments are still contained in the electrophoresis gel, but the results are usually not very good, as the gel matrix causes a lot of spurious background hybridization that obscures the specific hybridization signal. Instead the DNA bands in the agarose gel are transferred to a nitrocellulose or nylon membrane, providing a much 'cleaner' environment for the hybridization experiment.

Transfer of DNA bands from an agarose gel to a membrane makes use of the technique perfected in 1975 by Professor E.M. Southern and referred to as **Southern transfer**. The membrane is placed on the gel, and buffer allowed to soak through, carrying the DNA from the gel to the membrane where the DNA is bound. Sophisticated pieces of apparatus can be purchased to assist this process, but many molecular biologists prefer a homemade set-up incorporating a lot of paper towels and considerable balancing skills (Figure 10.1(b)). The same method can also be used for the transfer of RNA molecules (**'northern' transfer**) or proteins (**'western' transfer**). So far no one has come up with 'eastern' transfers.

Southern transfer results in a membrane that carries a replica of the DNA bands from the agarose gel. If the labelled probe is now applied, hybridization occurs and autoradiography (or the equivalent detection system for a non-radioactive probe) reveals which restriction fragment contains the cloned gene (Figure 10.1(c)). It is then possible to position the kanamycin resistance gene on the R6-5 restriction map (Figure 10.1(d)).

Southern transfer and hybridization can be used to locate the position of a cloned gene, or one isolated by PCR, within any DNA molecule for which a restriction map has been obtained. Note that this DNA molecule could itself be a recombinant plasmid or phage, with Southern hybridization used to determine the exact position of a gene within the cloned fragment. This is important as often the cloned DNA fragment is relatively large (e.g. 40 kb for a cosmid vector), whereas the gene of interest, contained somewhere in the cloned fragment, may be less than 1 kb in size. Also the cloned fragment may carry a number of genes in addition to the one under study. The strategies described in Chapter 8 for identifying a clone from a genomic library can therefore be followed up by Southern analysis of the recombinant DNA molecule to locate the precise position within the cloned DNA fragment of the gene being sought (Figure 10.2).

Figure 10.1 Southern
hybridization.

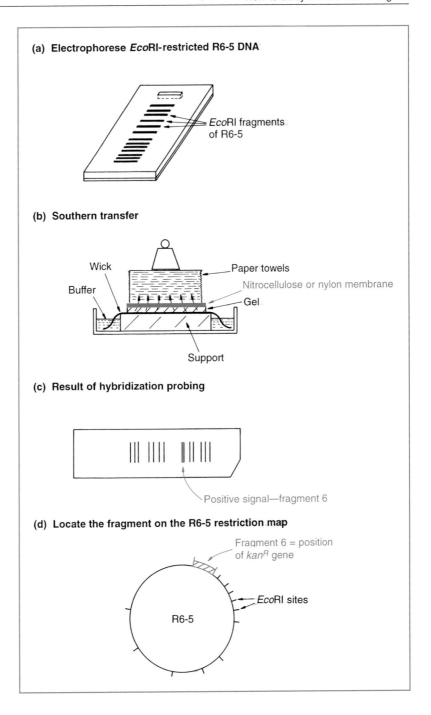

(a) **Electrophorese *Eco*RI-restricted R6-5 DNA**

*Eco*RI fragments
of R6-5

(b) **Southern transfer**

Wick
Buffer
Paper towels
Nitrocellulose or nylon membrane
Gel
Support

(c) **Result of hybridization probing**

Positive signal—fragment 6

(d) **Locate the fragment on the R6-5 restriction map**

Fragment 6 = position
of *kan^R* gene

*Eco*RI sites

R6-5

10.1.2

Locating the position of a gene on a large DNA molecule

Southern hybridization is feasible only if a restriction map can be worked out for the DNA molecule being studied. This means that the procedure is appropriate for most plasmids, bacteriophages and viruses, but cannot be

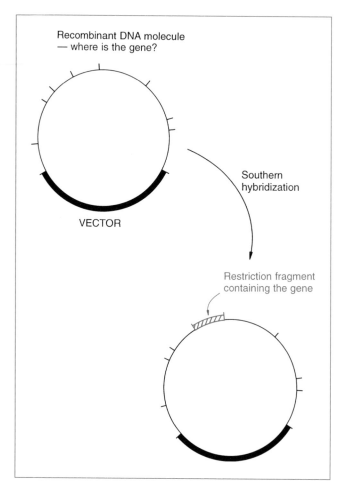

Figure 10.2
Southern hybridization can be used to locate the position of a cloned gene within a recombinant DNA molecule.

Recombinant DNA molecule
— where is the gene?

Southern hybridization

VECTOR

Restriction fragment containing the gene

used to locate genes on larger DNA molecules. Restriction mapping becomes very complicated with molecules more than about 250 kb in size, as can be appreciated by referring back to Figure 4.18. This example of restriction mapping is fairly straightforward as the λ molecule is not very big. Imagine how much more complicated the analysis would be if there were five times as many restriction sites. Other techniques must therefore be used to locate the positions of eukaryotic genes on chromosomal DNA molecules.

Separating chromosomes by gel electrophoresis

The first question to ask is, which chromosome carries the gene of interest? For some organisms this can be answered by a special type of Southern hybridization, involving not restriction fragments but intact chromosomal DNA molecules, separated by a novel type of gel electrophoresis.

In conventional gel electrophoresis, as described on p. 68, the electric field is orientated along the length of the gel and the DNA molecules migrate in a straight line towards the positive pole (Figure 10.3(a)). Different sized

Figure 10.3 Conventional agarose gel electrophoresis and its limitations.

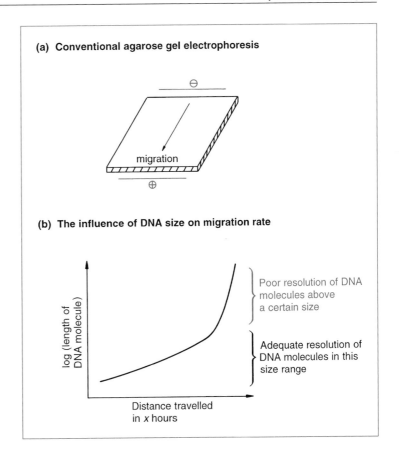

(a) Conventional agarose gel electrophoresis

migration

(b) The influence of DNA size on migration rate

log (length of DNA molecule)

Poor resolution of DNA molecules above a certain size

Adequate resolution of DNA molecules in this size range

Distance travelled in x hours

molecules can be separated because of the different rates at which they are able to migrate through the network of pores that make up the gel. However, only molecules within a certain size range can be separated in this way, because the difference in migration rate becomes increasingly small for larger molecules (Figure 10.3(b)). In practice, molecules larger than about 50 kb cannot be resolved efficiently by standard gel electrophoresis.

The limitations of standard gel electrophoresis can be overcome if a more complex electric field is used. Several different systems have been designed, but the principle is best illustrated by **orthogonal field alternation gel electrophoresis (OFAGE)**. Instead of being applied directly along the length of the gel, the electric field now alternates between two pairs of electrodes, each pair set at an angle of 45° to the length of the gel (Figure 10.4(a)). The result is a pulsed field, with the DNA molecules in the gel having continually to change direction in accordance with the pulses.

As the two fields alternate in a regular fashion the net movement of the DNA molecules in the gel is still from one end to the other, in more or less a straight line (Figure 10.4(a)). However, with every change in field direction each DNA molecule has to realign through 90° before its migration can continue. This is the key point, because a short molecule can realign faster than

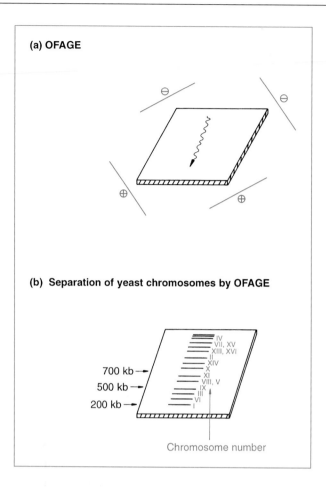

(a) OFAGE

(b) Separation of yeast chromosomes by OFAGE

700 kb →
500 kb →
200 kb →

IV
VII, XV
XIII, XVI
II
XIV
X
XI
VIII, V
IX
III
VI
I

Chromosome number

Figure 10.4
Orthogonal field alternation gel electrophoresis (OFAGE).

a long one, allowing the short molecule to progress towards the bottom of the gel more quickly. This added dimension increases the resolving power of the gel quite dramatically, so that molecules up to several thousand kilobases long can be separated. This size range includes the chromosomal molecules of many eukaryotes, including yeast, several important filamentous fungi, and protozoans such as the malaria parasite *Plasmodium falciparum*. Gels showing the chromosomes of these organisms can therefore be obtained (Figure 10.4(b)).

Orthogonal field alteration gel electrophoresis and related techniques such as **contour clamped homogeneous electric fields (CHEF)** and **field inversion gel electrophoresis (FIGE)** are important for a number of reasons. For example, the DNA from individual chromosomes can be purified from the gel, enabling a series of chromosomal gene libraries to be prepared. Each of these libraries, containing the genes from just one chromosome, is substantially smaller and easier to handle than a complete genomic library. In addition, chromosomal DNA molecules can be immobilized on a nitrocellulose or nylon membrane by Southern transfer and studied by hybridization analysis. In this way the chromosome that carries a cloned gene or one isolated by PCR can be identified.

In situ hybridization to visualize the position of a gene on a eukaryotic chromosome

Non-conventional gel electrophoresis techniques, including OFAGE, are at present limited to lower eukaryotes whose chromosomes are relatively small. The much larger molecules (>50 000 kb) of mammals and other higher eukaryotes are still some way beyond the capability of the current technology. Gene location on these larger DNA molecules can, however, be achieved by *in situ* **hybridization**, which has the added advantage of not only identifying which chromosome a gene lies on, but also providing information on the position of the gene within its chromosome.

In situ hybridization derives from the standard light microscopy techniques used to observe chromosomes in cells that are in the process of division (Figure 10.5(a)). With many organisms, individual chromosomes can be recognized by their shape and by the banding pattern produced by various types of stain. *In situ* hybridization provides a direct visual localization of a cloned gene on the light microscopic image of a chromosome.

Cells are treated with a fixative, attached to a glass slide and then incubated with ribonuclease and sodium hydroxide to degrade RNA and denature the DNA molecules. Base pairing between the individual polynucleotide strands is broken down, and the chromosomes unpack to a certain extent, exposing segments of DNA normally enclosed within their structure (Figure 10.5(b)). A sample of the gene is then labelled and applied to the chromosome preparation. Hybridization occurs between the gene and its chromosomal copy, resulting in a dark spot on an autoradiograph. The position of this spot indicates the location of the gene on its chromosome. Although a difficult technique, *in situ* hybridization with radioactively labelled probes has been used to position a number of genes on the human cytogenetic map.

As an alternative to radioactive labelling, a fluorescent marker can be attached to the probe and hybridization observed directly, using a special type of light microscope. If different fluorochromes are used, two or more genes can be probed at the same time, the different hybridization signals being distinguished by the distinctive colours of their fluorescences. This technique, **fluorescence *in situ* hybridization (FISH)**, is also frequently used with probes whose normal chromosomal locations are already known. This is particularly useful for studying cells in which chromosomal rearrangements have occurred. Rearrangements such as chromosome duplications, or the translocation of a segment of one chromosome to another, can be typed relatively quickly by FISH, more quickly than by conventional staining techniques.

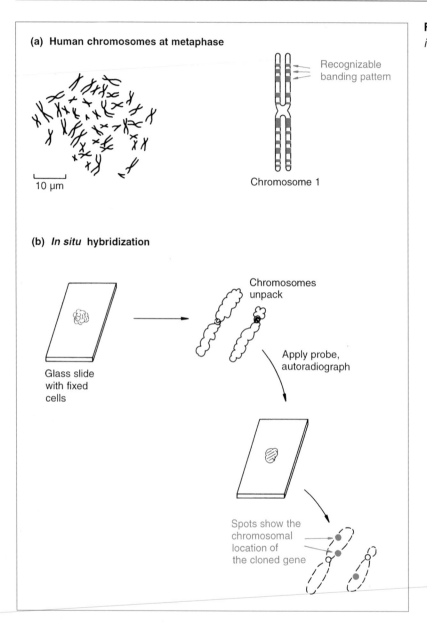

(a) **Human chromosomes at metaphase**

Recognizable banding pattern

10 µm

Chromosome 1

(b) *In situ* hybridization

Chromosomes unpack

Glass slide with fixed cells

Apply probe, autoradiograph

Spots show the chromosomal location of the cloned gene

Figure 10.5 Chromosomes and *in situ* hybridization.

10.2   DNA sequencing – working out the structure of a gene

Probably the most important technique available to the molecular biologist is DNA sequencing, by which the precise order of nucleotides in a piece of DNA can be determined. DNA sequencing methods have been around for 35 years, but only since the late 1970s has rapid and efficient sequencing been possible.

Two different techniques were developed almost simultaneously – the chain termination method by F. Sanger and A.R. Coulson in the UK, and the chemical degradation method by A. Maxam and W. Gilbert in the USA. The two techniques are radically different but equally valuable. Both allow DNA sequences of several kilobases in length to be determined in the minimum of time. The DNA sequence is now the first and most basic type of information to be obtained about a cloned gene.

10.2.1 The Sanger–Coulson method – chain-terminating nucleotides

The chain termination method requires single-stranded DNA and so the molecule to be sequenced is usually cloned into an M13 vector. This is because chain termination sequencing involves the enzymatic synthesis of a second strand of DNA, complementary to an existing template.

The primer

The first step in a chain termination sequencing experiment is to anneal a short oligonucleotide primer onto the recombinant M13 molecule (Figure 10.6(a)). This primer acts as the starting point for the complementary strand synthesis reaction carried out by the Klenow fragment of DNA polymerase I, or a related enzyme such as 'Sequenase', a modified version of the DNA polymerase encoded by bacteriophage T7. Remember that these enzymes need a double-stranded region from which to begin strand synthesis (p. 57). The primer anneals to the vector at a position adjacent to the polylinker.

Synthesis of the complementary strand

The strand synthesis reaction is started by adding the enzyme plus each of the four deoxynucleotides (2'-deoxyadenosine 5'-triphosphate (dATP), 2'-deoxythymidine 5'-triphosphate (dTTP), 2'-deoxyguanosine 5'-triphosphate (dGTP), 2'-deoxycytidine 5'-triphosphate (dCTP)). In addition a single modified nucleotide is also included in the reaction mixture. This is a **dideoxynucleotide** (e.g. dideoxyATP) which can be incorporated into the growing polynucleotide strand just as efficiently as the normal nucleotide, but which blocks further strand synthesis. This is because the dideoxynucleotide lacks the hydroxyl group at the 3' position of the sugar component (Figure 10.6(b)). This group is needed for the next nucleotide to be attached; chain termination therefore occurs whenever a dideoxynucleotide is incorporated by the enzyme.

If dideoxyATP is added to the reaction mix termination occurs at positions opposite thymidines in the template (Figure 10.6(c)). But termination does not always occur at the first T as normal dATP is also present and may be incorporated instead of the dideoxynucleotide. The ratio of dATP to dideoxyATP is such that an individual strand can be polymerized for a considerable distance before a dideoxyATP molecule is added. The result is that

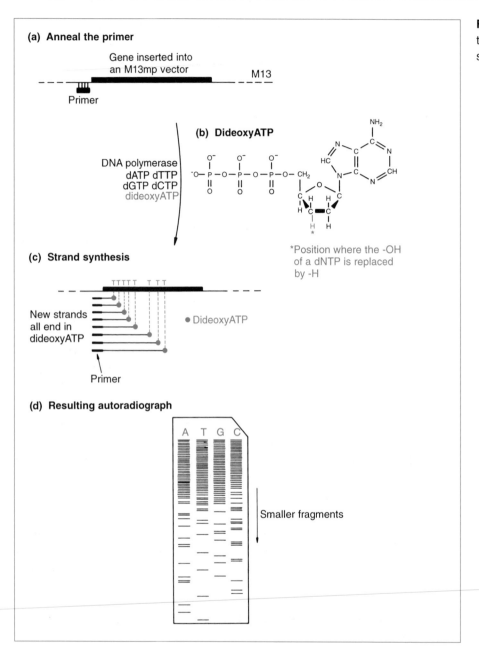

(a) Anneal the primer

Gene inserted into
an M13mp vector

M13

Primer

(b) DideoxyATP

NH₂

DNA polymerase
dATP dTTP
dGTP dCTP
dideoxyATP

O⁻ O⁻ O⁻
⁻O—P—O—P—O—P—O—CH₂
‖ ‖ ‖
O O O

*Position where the -OH
of a dNTP is replaced
by -H

(c) Strand synthesis

New strands
all end in
dideoxyATP

● DideoxyATP

Primer

(d) Resulting autoradiograph

A T G C

Smaller fragments

Figure 10.6 Chain termination DNA sequencing.

a family of new strands is obtained, all of different lengths, but each ending in dideoxyATP.

Four separate reactions result in four families of terminated strands

The strand synthesis reaction is carried out four times in parallel. As well as the reaction with dideoxyATP, there is one with dideoxyTTP, one with dideoxyGTP, and one with dideoxyCTP. The result is four distinct families of newly synthesized polynucleotides, one family containing strands all ending in dideoxyATP, one of strands ending in dideoxyTTP, etc.

The next step is to separate the components of each family so the lengths of each strand can be determined. This can be achieved by gel electrophoresis, although the conditions have to be carefully controlled as it is necessary to separate strands that differ in length by just one nucleotide. In practice, the electrophoresis is carried out in very thin polyacrylamide gels (less than 0.5 mm thick). The gels contain urea, which denatures the DNA so the newly synthesized strands dissociate from the templates. In addition, the electrophoresis is carried out at a high voltage, so the gel heats up to 60°C and above, making sure the strands do not reassociate in any way.

Each band in the gel contains only a small amount of DNA, so autoradiography has to be used to visualize the results (Figure 10.6(d)). The label is introduced into the new strands by including a radioactive deoxynucleotide (e.g. ^{32}P- or ^{35}S-dATP) in the reaction mixture for the strand synthesis step earlier in the experiment.

Reading the DNA sequence from the autoradiograph

Reading the sequence is very easy (Figure 10.7). First the band that has moved the furthest is located. This represents the smallest piece of DNA, the strand terminated by incorporation of the dideoxynucleotide at the first position in the template. The track in which this band occurs is noted. Let us say it is track A; the first nucleotide in the sequence is therefore A.

The next most mobile band corresponds to a DNA molecule one nucleotide longer than the first. The track is noted, T in the example shown in Figure 10.7; the second nucleotide is therefore T and the sequence so far is AT.

The process is continued along the autoradiograph until the individual bands become so bunched up that they cannot be separated from one another. Generally it is possible to read a sequence of about 400 nucleotides from one autoradiograph.

10.2.2   The Maxam–Gilbert method – chemical degradation of DNA

There are only a few similarities between the Sanger–Coulson and Maxam–Gilbert methods of DNA sequencing. The Maxam–Gilbert method requires double-stranded DNA fragments, so cloning into an M13 vector is not an essential first step. Neither is a primer needed, because the basis of the Maxam–Gilbert technique is not synthesis of a new strand, but cleavage of the existing DNA molecule using chemical reagents that act specifically at a particular nucleotide.

There are several variations of the Maxam–Gilbert method, differing in details such as the way in which the labelled DNA is obtained, and the exact

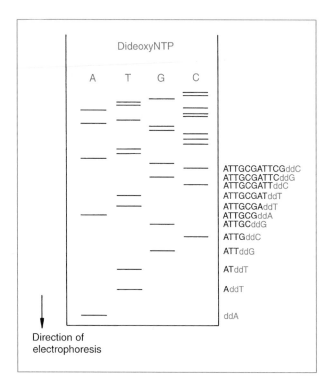

Figure 10.7 Interpreting the autoradiograph produced by a chain termination sequencing experiment. Each track contains the fragments produced by strand synthesis in the presence of one of the four dideoxynucleotide triphosphates (dideoxyNTPs). The sequence is read by identifying the track that each fragment lies in, starting with the one that has moved the furthest, and gradually progressing up through the autoradiograph.

nature of the cleavage reagents that are used. Most of these reagents are very toxic – chemicals that cleave DNA molecules in the test tube will do the same in the body, and great care must be taken when using them.

The following is a popular version of the Maxam–Gilbert technique. The double-stranded DNA fragment to be sequenced is first labelled by attaching a radioactive phosphorus group to the 5′ end of each strand (Figure 10.8(a)). Dimethylsulphoxide (DMSO) is then added and the labelled DNA sample heated to 90°C. This results in breakdown of the base pairing and dissociation of the DNA molecule into its two component strands. The two strands are separated from one another by gel electrophoresis (Figure 10.8(b)), which works on the basis that one of the strands probably contains more purine nucleotides than the other and will therefore be slightly heavier and will run more slowly during the electrophoresis. One strand is purified from the gel and divided into four samples, each of which is treated with one of the cleavage reagents. In fact, the first set of reagents to be added cause a chemical modification in the nucleotides they are specific for, making the strand susceptible to cleavage at that nucleotide when an additional chemical – piperidine – is added (Figure 10.8(c)). The modification and cleavage reactions are carried out under conditions that result in only one breakage per strand.

Some of the cleaved fragments retain the ^{32}P label at their 5′ ends. After electrophoresis, using the same special conditions as for chain termination

Figure 10.8 One version of DNA sequencing by the chemical degradation method.

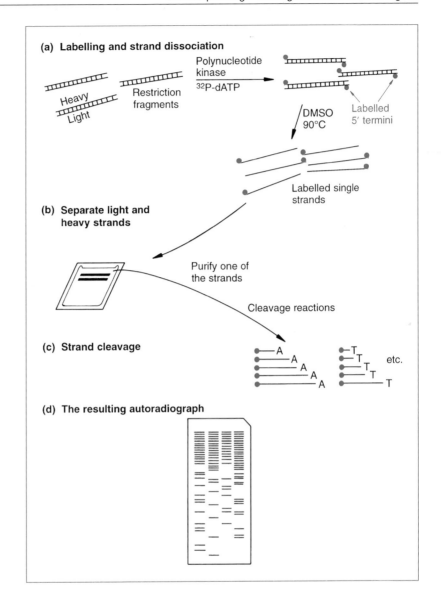

(a) **Labelling and strand dissociation**

Heavy

Light

Restriction fragments

Polynucleotide kinase

^{32}P-dATP

DMSO 90°C

Labelled 5′ termini

Labelled single strands

(b) **Separate light and heavy strands**

Purify one of the strands

Cleavage reactions

(c) **Strand cleavage**

A A A A A

T T T T T

etc.

(d) **The resulting autoradiograph**

sequencing, the bands visualized by autoradiography represent these labelled fragments. The nucleotide sequence can now be read from the autoradiograph exactly as for a chain termination experiment (Figure 10.8(d)).

10.2.3

Sequencing PCR products

As described on p. 187, some PCR experiments are designed so that the information that is required about the gene or other DNA sequence being studied can be obtained simply by examining the products by gel electrophoresis. Often, however, it is necessary to determine the sequence of the amplified DNA fragment. This can be achieved by cloning the PCR product

(p. 188) and using a standard chain termination or chemical degradation sequencing method as described above.

One problem with this approach is that the sequences of individual clones might not faithfully represent the sequence of the original template DNA molecule, because of the errors that are occasionally introduced by *Taq* polymerase during the amplification process (p. 192). Methods that sequence a PCR product directly, without the need for cloning, are therefore required.

Direct sequencing of PCR products

The direct sequencing method is based on the Sanger–Coulson technique and therefore requires single-stranded DNA as the starting material. The PCR product is of course double-stranded, so some means of purifying single strands is needed. There are several possibilities, the best of these being to carry out the initial PCR with one normal and one modified primer, the modified primer altered in such a way that the DNA strands synthesized from it are easily purified. A clever way of doing this is by attaching small magnetic beads to one of the primers. After the PCR, single-stranded DNA is obtained by separating the 'magnetic' strand from the ordinary strand (Figure 10.9). A similar method makes use of a biotin-labelled primer, with the single strands separated by binding to avidin, a protein that has a high affinity for biotin (p. 168).

Once single-stranded DNA has been purified, the remainder of the procedure is similar to the standard Sanger–Coulson method, in which families of chain-terminated molecules are synthesized by a DNA polymerase such as Sequenase. The only complication concerns the sequencing primer used as the starting point for the strand synthesis reactions. In the standard sequencing procedure this primer anneals to a site within the M13 vector, adjacent to the polylinker into which the DNA to be sequenced is cloned (Figure 10.6). This 'universal' primer cannot be used with PCR products as these do not contain the appropriate M13 sequences. Instead one of the primers used for the initial PCR is also used to prime the sequencing reactions. This primer has to be complementary to the purified single strands, and therefore is the primer that was not labelled with magnetic beads or with biotin.

Thermal cycle sequencing

Is it always necessary to purify single strands in order to sequence a PCR product? The answer is no, because it is possible to combine PCR and DNA sequencing into a single reaction, resulting in the technique called **thermal cycle sequencing**.

In thermal cycle sequencing the reaction mixture that is prepared is similar to a PCR mixture but with two important exceptions. The first is that only one primer is added to the reaction, which means that the amplification process characteristic of PCR cannot occur, and instead all that happens is

Figure 10.9 One way of purifying single-stranded DNA from a double-stranded PCR product. One of the primers is labelled with a magnetic bead. After PCR, the double-stranded products are denatured and the magnetic strands separated from the unlabelled strands.

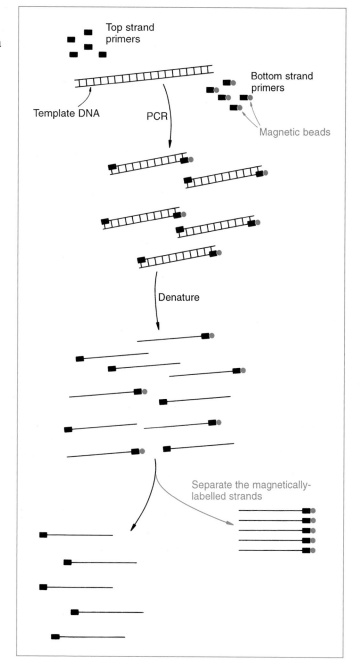

that one strand of the template DNA is copied. However, this copying is repeated many times because the reaction is thermally cycled, exactly as in a real PCR, and the enzyme that does the copying is a thermostable one such as *Taq* polymerase.

The second difference between a thermal cycle sequencing reaction and a PCR is that the former is carried out four times in parallel, each time with

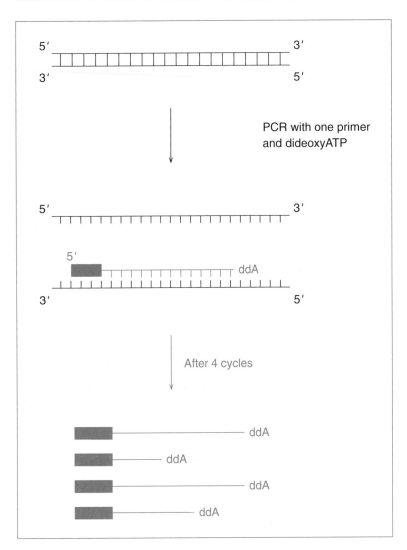

5′   3′

3′   5′

PCR with one primer
and dideoxyATP

5′   3′

5′

ddA

3′   5′

After 4 cycles

ddA

ddA

ddA

ddA

Figure 10.10 The basis to thermal cycle sequencing. A PCR is set up with just one primer and one of the dideoxyNTPs. One of the template strands is copied into a family of chain-terminated polynucleotides. ddA = dideoxyATP.

a different dideoxynucleotide included in the reaction. The new molecules that are synthesized are therefore chain-terminated (Figure 10.10), and the sequence of the template molecule can be read by running the products of the four reactions in a polyacrylamide gel, in the same way as for a standard Sanger–Coulson experiment.

10.2.4   Automated DNA sequencing

In the traditional methodology for Sanger–Coulson DNA sequencing, the chain-terminated molecules that are synthesized are radioactively labelled and the DNA sequence is read from an autoradiograph. This is not the only labelling and detection strategy that can be used with the chain termination method. As with *in situ* hybridization (p. 203), radioactive labelling has, in recent years, been supplemented by the use of fluorescent labels, leading to new horizons for DNA sequencing.

Fluorescent labels are usually attached to the dideoxynucleotides, so each chain-terminated molecule caries a single label at its 3′ end (Figure 10.11(a)). A different fluorochrome can be used for each of the four dideoxyNTPs, which has the major advantage that it is no longer necessary to carry out the four reactions in separate tubes. Now it is possible to perform a single sequencing reaction with all four dideoxyNTPs, because molecules terminated with different dideoxyNTPs can be identified from their distinctive fluorescent signals.

How are the fluorescent signals detected? A special type of imaging system is needed, one that involves the use of a computer to read the DNA sequence rather than relying on the eyes of a molecular biologist. The reaction products are loaded into a single well of a polyacrylamide gel or, in the latest technology, into a single tube of a capillary electrophoresis system, and then run past the fluorescence detector (Figure 10.11(b)). The detector identifies the fluorescent signal emitted by each band and feeds the data to the computer which converts the information into the appropriate DNA sequence. The sequence can be printed out for examination by the operator (Figure 10.11(c)), or entered directly into a storage device for future analysis. Automated sequencers can read up to 96 different sequences in a 2-hour period, and hence can acquire data much more rapidly than is possible by manual sequencing.

10.2.5 Building up a long DNA sequence

A single chain termination experiment carried out by hand produces about 400 nucleotides of sequence, and a single run in an automated sequencer gives about 750 nucleotides. But most genes are much longer than this; how can a sequence of several kilobases be obtained?

The answer is to perform DNA sequencing experiments with a lot of different cloned fragments or PCR products, all derived from a single larger DNA molecule (Figure 10.12). These fragments should overlap, so the individual DNA sequences will themselves overlap. The overlaps can be located, either by eye or using a computer, and the master sequence or **contig** (an abbreviation of contiguous sequence) gradually built up.

There are several ways of producing the overlapping fragments. Prior to cloning, the DNA molecule could be cleaved with two different restriction endonucleases, producing one set of fragments with, say, Sau3A and another with AluI (Figure 10.13). This was a popular method of producing overlapping sequences but suffers from the drawback that the restriction sites might be inconveniently placed and individual fragments may be too long to be completely sequenced. Often four or five different restriction endonucleases have to be used to clear up all the gaps in the master sequence.

An alternative is to break up the DNA molecule by **sonication**, which

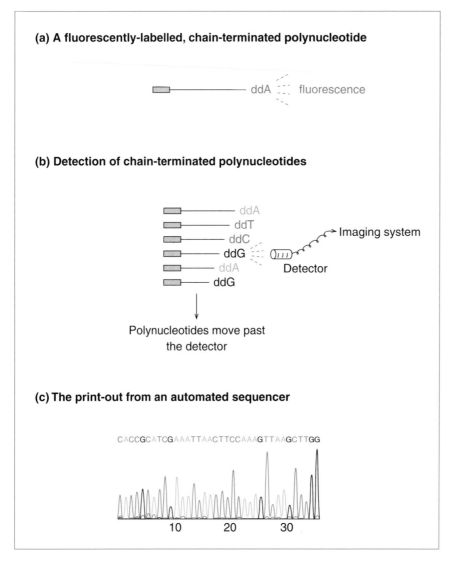

Figure 10.11 Automated DNA sequencing. (a) For automated sequencing, each dideoxyNTP is labelled with a fluorescent marker. (b) Each dideoxyNTP is labelled with a different fluorochrome, so the chain-terminated polynucleotides are distinguished as they pass by the detector. (c) An example of a sequence printout.

cleaves the DNA in a more random fashion and so give a greater possibility that the resulting fragments will overlap and a contiguous master sequence will be obtained. The fragments resulting from sonication have a variety of 3′ and 5′ overhangs but these can be converted into blunt ends by treatment with the appropriate enzymes prior to cloning in the normal way.

10.2.6   The achievements of DNA sequencing

The first DNA molecule to be completely sequenced was the 5386 nucleotide genome of bacteriophage φX174, which was completed in 1975. This was quickly followed by sequences for SV40 virus (5243 bp) in 1977 and pBR322 (4363 bp) in 1978. Gradually sequencing was applied to larger molecules. Professor Sanger's group published the sequence of the human mitochondrial

Figure 10.12 Building up a long DNA sequence from a series of short overlapping ones.

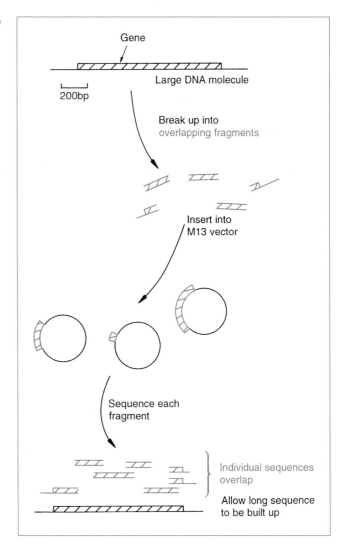

Gene

Large DNA molecule

200bp

Break up into
overlapping fragments

Insert into
M13 vector

Sequence each
fragment

Individual sequences
overlap

Allow long sequence
to be built up

Figure 10.13 Building up a long DNA sequence by determining the sequences of overlapping restriction fragments.

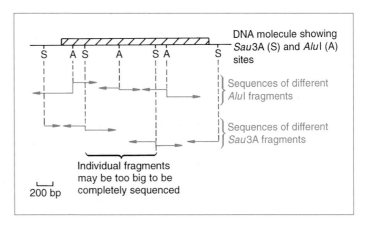

DNA molecule showing
Sau3A (S) and AluI (A)
sites

Sequences of different
AluI fragments

Sequences of different
Sau3A fragments

Individual fragments
may be too big to be
completely sequenced

200 bp

genome (16.6 kb) in 1981 and of bacteriophage λ (49 kb) in 1982. Nowadays sequences of 100–200 kb are routine and most research laboratories have the necessary expertise to generate this amount of information.

The pioneering projects today are the massive genome initiatives, each aimed at obtaining the nucleotide sequence of the entire genome of a particular organism. The first chromosome sequence, for chromosome III of the yeast *Saccharomyces cerevisiae*, was published in 1992, and the entire yeast genome was completed in 1996. There are now complete genome sequences for the worm *Caenorhabditis elegans*, the fly *Drosophila melanogaster*, the plant *Arabidopsis thaliana* and the human *Homo sapiens*, not to mention over 100 different microbial species. We will examine these genome projects in more detail in Chapter 12.

Further reading

Carle, G.E. & Olson, M.V. (1985) An electrophoretic karyotype for yeast. *Proceedings of the National Academy of Sciences of the USA*, **82**, 3756–60. An application of OFAGE.

Heiskanen, M., Peltonen, L. & Palotie, A. (1996) Visual mapping by high resolution FISH. *Trends in Genetics*, **12**, 379–82.

Maxam, A. & Gilbert, W. (1977) A new method of sequencing DNA. *Proceedings of the National Academy of Sciences of the USA*, **74**, 560–64.

Oliver, S.G., van der Hart, O.J.M., Agostine Carbone, M.T. *et al.* (1992) The complete DNA sequence of yeast chromosome III. *Nature*, **357**, 38–46.

Prober, J.M., Trainor, G.L., Dam, R.J. *et al.* (1987) A system for rapid DNA sequencing with fluorescent chain-terminating dideoxynucleotides. *Science*, **238**, 336–41.

Sanger, F., Nicklen, S. & Coulson, A.R. (1977) DNA sequencing with chain-terminating inhibitors. *Proceedings of the National Academy of Sciences of the USA*, **74**, 5463–7.

Southern, E.M. (1975) Detection of specific sequences among DNA fragments separated by gel electrophoresis. *Journal of Molecular Biology*, **98**, 503–7. [Southern hybridization.]

The *Arabidopsis* Genome Initiative (2000) Analysis of the genome sequence of the flowering plant *Arabidopsis thaliana*. *Nature*, **408**, 796–815.

Chapter 11 Studying Gene Expression and Function

All genes have to be expressed in order to function. The first step in expression is transcription of the gene into a complementary RNA strand (Figure 11.1(a)). For some genes – for example those coding for transfer RNA (tRNA) and ribosomal RNA (rRNA) molecules – the transcript itself is the functionally important molecule. For other genes the transcript is translated into a protein molecule.

To understand how a gene is expressed, the RNA transcript must be studied. In particular, the molecular biologist will want to know whether the transcript is a faithful copy of the gene, or whether segments of the gene are missing from the transcript (Figure 11.1(b)). These missing pieces are called introns and considerable interest centres on their structure and possible function. In addition to introns, the exact locations of the start and end points of transcription are important. Most transcripts are copies not only of the gene itself, but also of the nucleotide regions either side of it (Figure 11.1(c)). The signals that determine the start and finish of the transcription process are only partly understood, but they must be located if the expression of a gene is to be studied.

In this chapter we will begin by looking at the methods used for **transcript analysis**. These methods can be used to determine if a gene contains introns and to map the positions of the start and end points for transcription. Then we will briefly consider a few of the numerous techniques developed in recent years for examining how expression of a gene is regulated. These techniques are important as aberrations in gene regulation underlie many clinical disorders. Finally, we will tackle the difficult problem of how to identify the translation product of a gene.

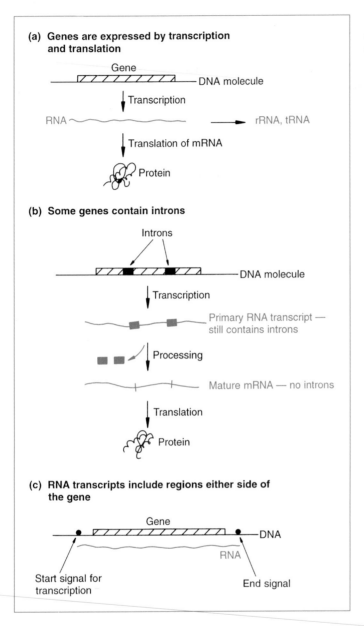

(a) Genes are expressed by transcription and translation

Gene

DNA molecule

Transcription

RNA — rRNA, tRNA

Translation of mRNA

Protein

(b) Some genes contain introns

Introns

DNA molecule

Transcription

Primary RNA transcript — still contains introns

Processing

Mature mRNA — no introns

Translation

Protein

(c) RNA transcripts include regions either side of the gene

Gene

DNA

RNA

Start signal for transcription

End signal

Figure 11.1 Some fundamentals of gene expression. mRNA = messenger RNA, tRNA = transfer RNA, rRNA = ribosomal RNA.

11.1 Studying the transcript of a cloned gene

Most methods of transcript analysis are based on hybridization between the RNA transcript and a fragment of DNA containing the relevant gene. Nucleic acid hybridization occurs just as readily between complementary DNA and RNA strands as between single-stranded DNA molecules. The resulting DNA–RNA hybrid can be analysed by electron microscopy or with single-strand-specific nucleases.

11.1.1 Electron microscopy of nucleic acid molecules

Electron microscopy can be used to visualize nucleic acid molecules, as long as the polynucleotides are first treated with chemicals that increase their apparent diameter. Untreated molecules are simply too thin to be seen.

Usually the DNA molecules are mixed with a protein such as cytochrome c which binds to the polynucleotides, coating the strands in a thick shell. The coated molecules must be stained with uranyl acetate or some other electron-dense material to enhance the appearance of the preparation (Figure 11.2). Quite spectacular views of nucleic acid molecules can be obtained.

In the past, electron microscopy has been used primarily to analyse hybridization between different DNA molecules, but in recent years the technique has become increasingly important in the study of DNA–RNA hybrids. It is particularly useful for determining if a gene contains introns. Consider the appearance of a hybrid between a DNA strand, containing a gene, and its RNA transcript. If the gene contains introns, these regions of the DNA strand will have no similarity with the RNA transcript and so cannot base pair. Instead they 'loop out', giving a characteristic appearance when observed with the electron microscope (Figure 11.3). The number and positions of these loops correspond directly to the number and positions of the

Figure 11.2 Preparing a DNA molecule for observation with the electron microscope.

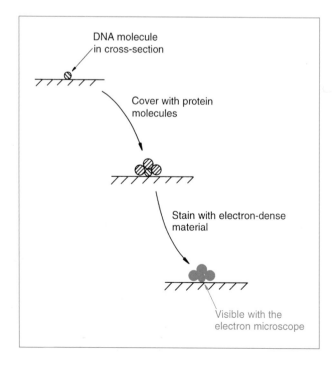

DNA molecule in cross-section

Cover with protein molecules

Stain with electron-dense material

Visible with the electron microscope

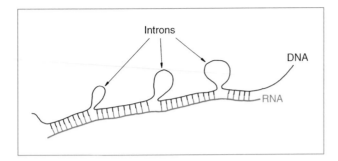

Figure 11.3 The appearance under the electron microscope of a DNA–RNA hybrid formed between a gene containing an intron and its processed transcript.

introns in the gene. Further information can then be obtained by sequencing the gene and looking for the characteristic features that mark the boundaries of introns. If a cDNA clone is available its sequence, which of course lacks the introns, can be compared with the gene sequence to locate the introns with precision.

11.1.2 Analysis of DNA–RNA hybrids by nuclease treatment

The second method for studying a DNA–RNA hybrid involves a single-strand-specific nuclease such as S1 (p. 54). This enzyme degrades single-stranded DNA or RNA polynucleotides, including single-stranded regions at the end of predominantly double-stranded molecules, but has no effect on double-stranded DNA or on DNA–RNA hybrids. If a DNA molecule containing a gene is hybridized to its RNA transcript, and then treated with S1 nuclease, the non-hybridized single-stranded DNA regions at each end of the hybrid are digested, along with any looped-out introns (Figure 11.4). The result is a completely double-stranded hybrid. The single-stranded DNA fragments protected from S1 nuclease digestion can be recovered if the RNA strand is degraded by treatment with alkali.

Unfortunately the manipulations shown in Figure 11.4 are not very informative. The sizes of the protected DNA fragments could be measured by gel electrophoresis, but this does not allow their order or relative positions to be determined. However, a few subtle modifications to the technique allow the precise start and end points of the transcript and of any introns it contains to be mapped onto the DNA sequence.

In the example shown in Figure 11.5, a *Sau*3A fragment that contains 100 bp of coding region, along with 300 bp of the leader sequence preceding the gene, has been cloned into an M13 vector and obtained as a single-stranded molecule. A sample of the RNA transcript is added and allowed to anneal to the DNA molecule. The DNA molecule is still primarily single-stranded but now has a small region protected by the RNA transcript. All but this protected region is digested by S1 nuclease and the RNA degraded with alkali, leaving a short single-stranded DNA fragment. If these manipulations are examined closely it will become clear that the size of this single-

Figure 11.4 The effect of S1 nuclease on a DNA–RNA hybrid.

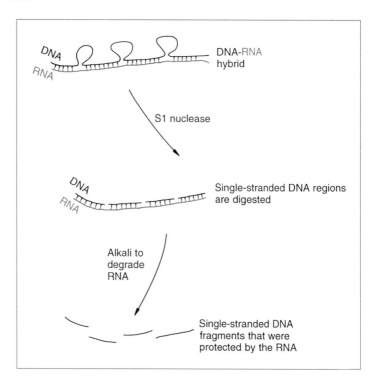

DNA
RNA

DNA-RNA hybrid

S1 nuclease

DNA
RNA

Single-stranded DNA regions are digested

Alkali to degrade RNA

Single-stranded DNA fragments that were protected by the RNA

stranded fragment corresponds to the distance between the transcription start point and the right-hand *Sau*3A site. The size of the single-stranded fragment is therefore determined by gel electrophoresis and this information is used to locate the transcription start point on the DNA sequence. Exactly the same strategy could locate the end point of transcription and the junction points between introns and exons. The only difference would be the position of the restriction site chosen to delimit one end of the protected single-stranded DNA fragment.

11.1.3   Transcript analysis by primer extension

S1 nuclease analysis is a powerful technique that allows both the 5′ and 3′ termini of a transcript, as well as the positions of intron–exon boundaries, to be identified. The final method of transcript analysis that we will consider – **primer extension** – is less adaptable because it can only identify the 5′ end of an RNA. It is, nonetheless, an important technique that is frequently used to check the results of S1 analyses.

Primer extension can only be used if at least part of the sequence of the transcript is known. This is because a short oligonucleotide primer must be annealed to the RNA at a known position, ideally within 100–200 nucleotides of the 5′ end of the transcript. Once annealed, the primer is extended by reverse transcriptase (p. 58), which continues copying the RNA strand until it reaches the 5′ terminus (Figure 11.6). The 3′ end of this newly synthesized

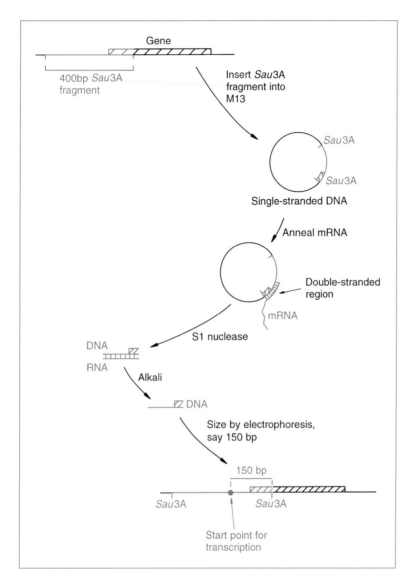

Figure 11.5 Locating a transcription start point by S1 nuclease mapping.

strand of DNA therefore corresponds with the 5′ terminus of the transcript. Locating the position of this terminus on the DNA sequence is achieved simply by determining the length of the single-stranded DNA molecule and correlating this information with the annealing position of the primer.

11.1.4 Other techniques for studying RNA transcripts

In recent years a broad range of RNA manipulative techniques have been developed and their use has resulted in a number of important advances in understanding how transcripts are synthesized and processed. The techniques include the following.

Figure 11.6 Locating a transcription start point by primer extension.

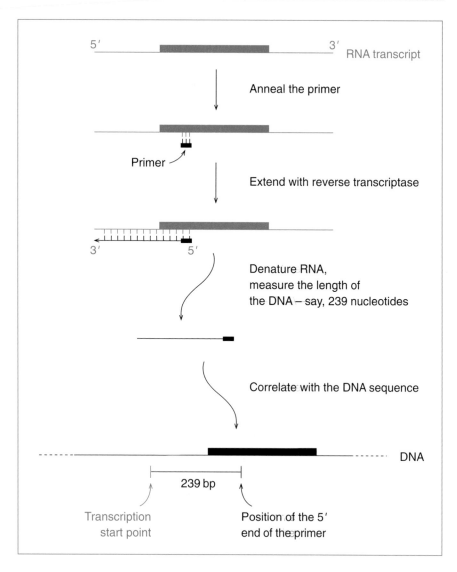

Northern hybridization

This technique is the RNA equivalent of Southern hybridization (p. 198). An RNA extract is electrophoresed in an agarose gel, using a denaturing electrophoresis buffer (e.g. one containing formaldehyde) to ensure that the RNAs do not form inter- or intramolecular base pairs, as base pairing would affect the rate at which the molecules migrate through the gel. After electrophoresis, the gel is blotted onto a nylon or nitrocellulose membrane, and hybridized with a labelled probe (Figure 11.7). If the probe is a cloned gene, the band that appears in the autoradiograph is the transcript of that gene. The size of the transcript can be determined from its position within the gel, and if RNA from different tissues is run in different lanes

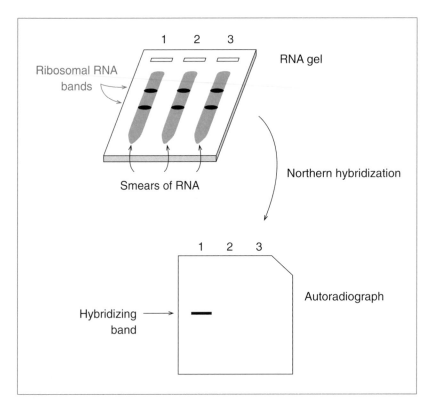

Figure 11.7 Northern hybridization. Three RNA extracts from different tissues have been electrophoresed in an agarose gel. The extracts are made up of many RNAs of different lengths so each gives a smear of RNA, but two distinct bands are seen, one for each of the abundant ribosomal RNAs. The sizes of these rRNAs are known (e.g. 4718 and 1874 nucleotides in mammals) so they can be used as internal size markers. The gel is transferred to a membrane, probed with a cloned gene, and the results visualized by autoradiography. Only lane 1 gives a band, showing that the cloned gene is expressed only in the tissue from which this RNA extract was obtained.

of the gel then the possibility that the gene is differentially expressed can be examined.

Reverse transcription–PCR (RT–PCR)

This technique enables RNA to be used as the template for a polymerase chain reaction (PCR). The first step in **RT–PCR** is to convert the RNA molecules into single-stranded cDNA with reverse transcriptase. Once this preliminary step has been carried out the PCR primers and *Taq* polymerase are added and the experiment proceeds exactly as in the standard technique (Figure 11.8). Some thermostable polymerases are able to make DNA copies of both RNA and DNA molecules (i.e. they have both reverse transcriptase and DNA-dependent DNA polymerase activities) and so can carry out all the steps of RT–PCR in a single reaction. The product of RT–PCR is many double-stranded complementary DNA (cDNA) copies of the RNA template, though probably not copies of the entire RNA molecule because usually the primer annealing sites lie within the transcript rather than at the extreme ends. Reverse transcription–PCR is often used to test RNA extracts from different tissues for the presence of a particular transcript in order to determine the expression pattern of a gene. A modified version called **rapid amplification of cDNA ends (RACE)** can be used to identify the 5′ and 3′ termini of RNA molecules.

Figure 11.8 Reverse
transcription–PCR (RT–PCR).

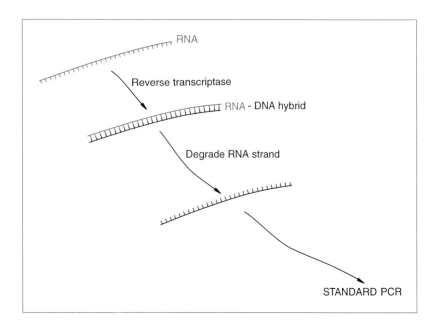

RNA sequencing

RNA sequencing is usually achieved by sequence analysis of the product of an RT–PCR. Direct methods for sequencing RNA molecules have been devised but these are inefficient and, importantly, the RNA molecule must be purified before it can be sequenced. It is possible to obtain pure samples of the RNA genomes of viruses, and of very abundant cellular RNAs such as ribosomal RNA molecules, but purification of a single mRNA is very difficult. However, if the primers for an RT–PCR are designed correctly, just the single, target mRNA is copied, and sequence analysis of the RT–PCR product provides the sequence of this mRNA.

11.2 Studying the regulation of gene expression

Few genes are expressed all the time. Most are subject to regulation and are switched on only when their gene product is required by the cell. The simplest gene regulation systems are found in bacteria such as *E. coli*, which can regulate expression of genes for biosynthetic and metabolic processes, so that gene products that are not needed are not synthesized. For instance, the genes coding for the enzymes involved in tryptophan biosynthesis can be switched off when there are abundant amounts of tryptophan in the cell, and switched on again when tryptophan levels drop. Similarly, genes for the utilization of sugars such as lactose are activated only when the relevant sugar is there to be metabolized. In higher organisms gene

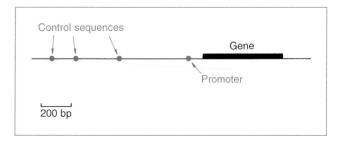

Figure 11.9 Possible positions for control sequences in the region upstream of a gene.

regulation is more complex because there are many more genes to control. Differentiation of cells involves wholesale changes in gene expression patterns, and the process of development from fertilized egg cell to adult requires coordination between different cells as well as time-dependent changes in gene expression.

Many of the problems in gene regulation require a classical genetic approach: genetics enables genes that control regulation to be distinguished, allows the biochemical signals that influence gene expression to be identified, and can explore the interactions between different genes and gene families. It is for this reason that most of the breakthroughs in understanding development in higher organisms have started with studies of the fruit fly *Drosophila melanogaster*. Gene cloning and DNA analysis complement classical genetics as they provide much more detailed information on the molecular events involved in regulating the expression of a single gene.

We now know that a gene subject to regulation has one or more control sequences in its upstream region (Figure 11.9) and that the gene is switched on and off by the attachment of regulatory proteins to these sequences. A regulatory protein may repress gene expression, in which case the gene is switched off when the protein is bound to the control sequence, or alternatively the protein may have a positive or enhancing role, switching on or increasing expression of the target gene. In this section we will examine methods for locating control sequences and determining their roles in regulating gene expression.

11.2.1 Identifying protein binding sites on a DNA molecule

A control sequence is a region of DNA that can bind a regulatory protein. It should therefore be possible to identify control sequences upstream of a gene by searching the relevant region for protein binding sites. There are three different ways of doing this.

Gel retardation of DNA–protein complexes

Proteins are quite substantial structures and a protein attached to a DNA molecule results in a large increase in overall molecular mass. If this increase can be detected, a DNA fragment containing a protein binding site will

Figure 11.10 A bound protein decreases the mobility of a DNA fragment during gel electrophoresis.

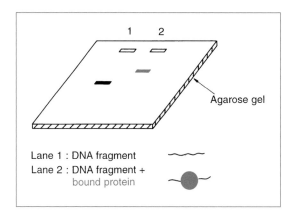

have been identified. In practice a DNA fragment carrying a bound protein is identified by gel electrophoresis, as it has a lower mobility than the uncomplexed DNA molecule (Figure 11.10). The procedure is referred to as **gel retardation**.

In a gel retardation experiment (Figure 11.11), the region of DNA upstream of the gene being studied is digested with a restriction endonuclease and then mixed with the regulatory protein or, if the protein has not yet been purified, with an unfractionated extract of nuclear protein (remember that gene regulation occurs in the nucleus). The restriction fragment containing the control sequence forms a complex with the regulatory protein: all the other fragments remain as 'naked' DNA. The location of the control sequence is then determined by finding the position on the restriction map of the fragment that is retarded during gel electrophoresis. The precision with which the control sequence can be located depends on how detailed the restriction map is and how conveniently placed the restriction sites are. A single control sequence may be less than 10 bp in size, so gel retardation is rarely able to pinpoint it exactly. More precise techniques are therefore needed to delineate the position of the control sequence within the fragment identified by gel retardation.

Footprinting with DNase I

The procedure generally called **footprinting** enables a control region to be positioned within a restriction fragment that has been identified by gel retardation. Footprinting works on the basis that the interaction with a regulatory protein protects the DNA in the region of a control sequence from the degradative action of an endonuclease such as deoxyribonuclease (DNase) I (Figure 11.12). This phenomenon can be used to locate the protein binding site on the DNA molecule.

The DNA fragment being studied is first labelled at one end with a radioactive marker, and then complexed with the regulatory protein (Figure

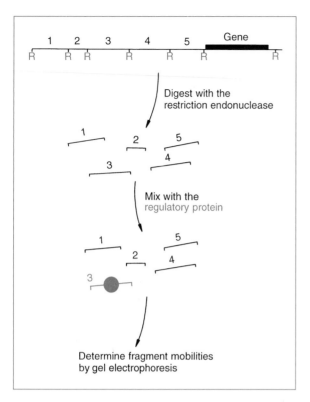

Figure 11.11 Carrying out a gel retardation experiment.

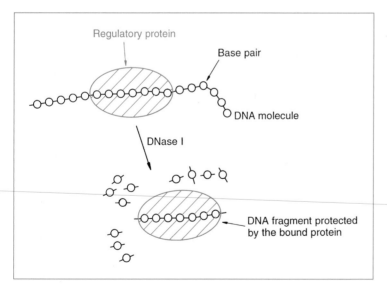

Figure 11.12 A bound protein protects a region of a DNA molecule from degradation by a nuclease such as DNase I.

11.13(a)). Then DNase I is added, but the amount used is limited so that complete degradation of the DNA fragment does not occur. Instead the aim is to cut each molecule at just a single phosphodiester bond (Figure 11.13(b)). If the DNA fragment has no protein attached to it the result of this treat-

Figure 11.13 DNase I footprinting.

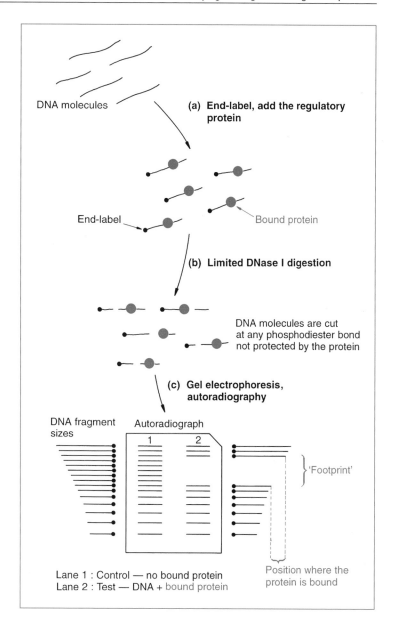

DNA molecules

(a) End-label, add the regulatory protein

End-label

Bound protein

(b) Limited DNase I digestion

DNA molecules are cut at any phosphodiester bond not protected by the protein

(c) Gel electrophoresis, autoradiography

DNA fragment sizes

Autoradiograph

1   2

'Footprint'

Position where the protein is bound

Lane 1 : Control — no bound protein
Lane 2 : Test — DNA + bound protein

ment is a family of labelled fragments, differing in size by just one nucleotide each.

After separation on a polyacrylamide gel, the family of fragments appears as a ladder of bands on an autoradiograph (Figure 11.13(c)). However, the bound protein protects certain phosphodiester bonds from being cut by DNase I, meaning that in this case the family of fragments is not complete, as the fragments resulting from cleavage within the control sequence are absent. Their absence shows up on the autoradiograph as a 'footprint', clearly seen in Figure 11.13(c). The region of the DNA molecule containing the control sequence can now be worked out from the sizes of the fragments on either side of the footprint.

Modification interference assays

Gel retardation analysis and footprinting enable control sequences to be located but do not give information about the interaction between the binding protein and the DNA molecule. The more precise of these two techniques – footprinting – only reveals the region of DNA that is protected by the bound protein. Proteins are relatively large compared with a DNA double helix, and can protect several tens of base pairs when bound to a control sequence that is just a few base pairs in length (Figure 11.14). Footprinting therefore does not delineate the control region itself, only the region within which it is located.

Nucleotides that actually form attachments with a bound protein can be identified by the **modification interference assay**. As in footprinting, the DNA fragments must first be labelled at one end. Then they are treated with a chemical that modifies specific nucleotides, an example being dimethyl sulphate, which attaches methyl groups to guanine nucleotides (Figure 11.15). This modification is carried out under limiting conditions so an average of just one nucleotide per DNA fragment is modified. Now the DNA is mixed with the protein extract. The key to the assay is that the binding protein will probably not attach to the DNA if one of the guanines within the control region is modified, because methylation of a nucleotide interferes with the specific chemical reaction that enables it to form an attachment with a protein.

How is the absence of protein binding monitored? If the DNA and protein mixture is examined by agarose gel electrophoresis two bands will be seen, one comprising the DNA–protein complex and one containing DNA

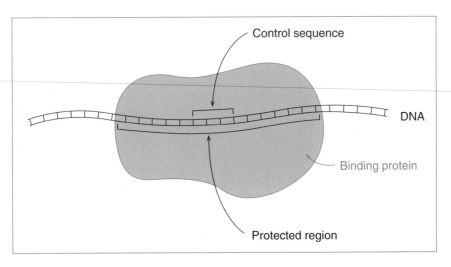

Control sequence

DNA

Binding protein

Protected region

Figure 11.14 A bound protein can protect a region of DNA that is much longer than the control sequence.

Figure 11.15 A
modification interference
assay.

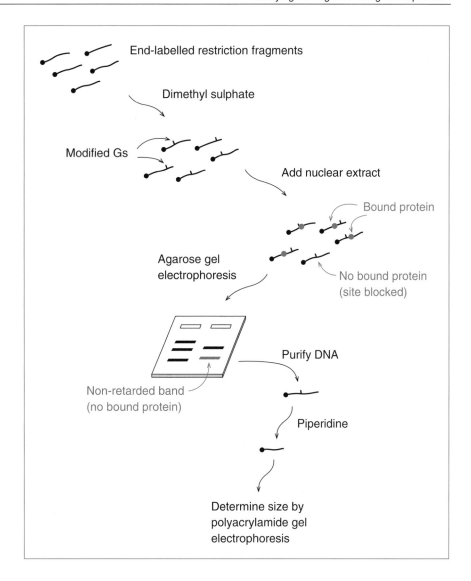

End-labelled restriction fragments

Dimethyl sulphate

Modified Gs

Add nuclear extract

Bound protein

Agarose gel
electrophoresis

No bound protein
(site blocked)

Non-retarded band
(no bound protein)

Purify DNA

Piperidine

Determine size by
polyacrylamide gel
electrophoresis

with no bound protein – in essence, this part of the procedure is a gel retardation assay (Figure 11.15). The band made up of unbound DNA is purified
from the gel and treated with piperidine, a chemical which cleaves DNA
molecules at methylated nucleotides (p. 208). The products of piperidine
treatment are now separated in a polyacrylamide gel and the results visualized by autoradiography. The size of the band or bands that appear on the
autoradiograph indicates the position in the DNA fragment of guanines
whose methylation prevented protein binding. These guanines lie within the
control sequence. The modification assay can now be repeated with chemicals that target A, T or C nucleotides to determine the precise position of the
control sequence.

11.2.2

Identifying control sequences by deletion analysis

Gel retardation, footprinting and modification interference assays are able to locate possible control sequences upstream of a gene, but they provide no information on the function of the individual sequences. **Deletion analysis** is a totally different approach that not only can locate control sequences (though only with the precision of gel retardation), but importantly also can indicate the function of each sequence.

The technique depends on the assumption that deletion of the control sequence will result in a change in the way in which expression of a cloned gene is regulated (Figure 11.16). For instance, deletion of a sequence that represses expression of a gene should result in that gene being expressed at a higher level. Similarly, tissue-specific control sequences can be identified as their deletion results in the target gene being expressed in tissues other than the correct one.

Reporter genes

Before carrying out a deletion analysis, a way must be found to assay the effect of a deletion on expression of the cloned gene. The effect will probably only be observed when the gene is cloned into the species it was originally obtained from: it will be no good assaying for light regulation of a plant gene if the gene is cloned in a bacterium.

Cloning vectors have now been developed for most organisms (Chapter 7), so cloning the gene under study back into its host should not cause a problem. The difficulty is that in most cases the host already pos-

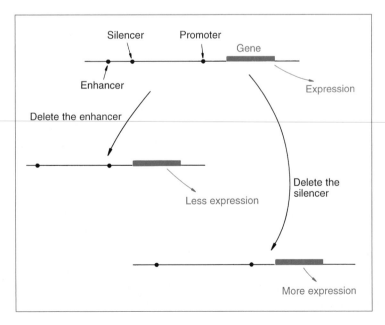

Figure 11.16 The principle behind deletion analysis.

Figure 11.17 A reporter gene.

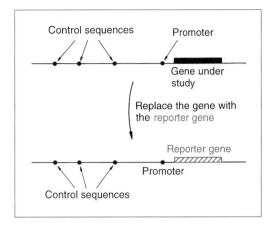

sesses a copy of the gene within its chromosomes. How can changes in the expression pattern of the cloned gene be distinguished from the normal pattern of expression displayed by the chromosomal copy of the gene? The answer is to use a **reporter gene**. This is a test gene that is fused to the upstream region of the cloned gene (Figure 11.17), replacing this gene. When cloned into the host organism the expression pattern of the reporter gene should exactly mimic that of the original gene, as the reporter gene is under the influence of exactly the same control sequences as the original gene.

The reporter gene must be chosen with care. The first criterion is that the reporter gene must code for a phenotype not already displayed by the host organism. The phenotype of the reporter gene must be relatively easy to detect after it has been cloned into the host, and ideally it should be possible to assay the phenotype quantitatively. These criteria have not proved difficult to meet and a variety of different reporter genes have been used in studies of gene regulation. A few examples are listed in Table 11.1.

Carrying out a deletion analysis

Once a reporter gene has been chosen and the necessary construction made, carrying out a deletion analysis is fairly straightforward. Deletions can be made in the upstream region of the construct by any one of several strategies, a simple example being shown in Figure 11.18. The effect of the deletion is then assessed by cloning the deleted construct into the host organism and determining the pattern and extent of expression of the reporter gene. An increase in expression will imply that a repressing or silencing sequence has been removed, a decrease will indicate removal of an activator or enhancer, and a change in tissue specificity (as shown in Figure 11.18) will pinpoint a tissue-responsive control sequence.

The results of a deletion analysis project have to be interpreted very carefully. Complications may arise if a single deletion removes two closely linked control sequences or, as is fairly common, two distinct control

Table 11.1 A few examples of reporter genes used in studies of gene regulation in higher organisms.

Gene[a]	Gene product	Assay
lacZ	β-galactosidase	Histochemical test
neo	Neomycin phosphotransferase	Kanamycin resistance
cat	Chloramphenicol acetyltransferase	Chloramphenicol resistance
dhfr	Dihydrofolate reductase	Methotrexate resistance
aphIV	Hygromycin resistance	Hygromycin resistance
lux	Luciferase	Bioluminescence
uidA	β-glucuronidase	Histochemical test

[a] All of these genes are obtained from *E. coli*, except for *lux* which has three sources: the luminescent bacteria *Vibrio harveyii* and *V. fischeri*, and the firefly *Photinus pyralis*.

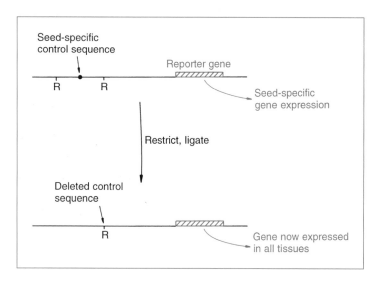

Figure 11.18 Deletion analysis. A reporter gene has been attached to the upstream region of a seed-specific gene from a plant. Removal of the restriction fragment bounded by the sites R deletes the control sequence that mediates seed-specific gene expression, so that the reporter gene is now expressed in all tissues of the plant.

sequences cooperate to produce a single response. Despite these potential difficulties deletion analyses, in combination with studies of protein binding sites, have provided important information about how the expression of individual genes is regulated, and have supplemented and extended the more broadly based genetic analyses of differentiation and development.

11.3 Identifying and studying the translation product of a cloned gene

In the last few years gene cloning has become increasingly useful in the study not only of genes themselves but also of the proteins coded by cloned genes.

Investigations into protein structure and function have benefited greatly from the development of techniques that allow mutations to be introduced at specific points in a cloned gene, resulting in directed changes in the structure of the encoded protein.

Before considering these procedures we should first look at the more mundane problem of how to isolate the protein coded by a cloned gene. In many cases this analysis will not be necessary, as the protein will have been characterized long before the gene cloning experiment is performed and pure samples of the protein will already be available. On the other hand, there are occasions when the translation product of a cloned gene has not been identified. A method for isolating the protein is then needed.

11.3.1 HRT and HART can identify the translation product of a cloned gene

Two related techniques, **hybrid-release translation (HRT)** and **hybrid-arrest translation (HART)**, are used to identify the translation product encoded by a cloned gene. Both depend on the ability of purified mRNA to direct synthesis of proteins in **cell-free translation systems**. These are cell extracts, usually prepared from germinating wheat seeds or from rabbit reticulocyte cells (both of which are exceptionally active in protein synthesis) and containing ribosomes, tRNAs and all the other molecules needed for protein synthesis. The mRNA sample is added to the cell-free translation system, along with a mixture of the 20 amino acids found in proteins, one of which is labelled (often ^{35}S-methionine is used). The mRNA molecules are translated into a mixture of radioactive proteins (Figure 11.19), which can be separated by gel electrophoresis and visualized by autoradiography. Each band represents a single protein coded by one of the mRNA molecules present in the sample.

Both HRT and HART work best when a cDNA clone prepared directly from the mRNA sample is available. For HRT the cDNA is denatured, immobilized on a nitrocellulose or nylon membrane, and incubated with the mRNA sample (Figure 11.20). The specific mRNA counterpart of the cDNA hybridizes and remains attached to the membrane. After discarding the unbound molecules, the hybridized mRNA is recovered and translated in a cell-free system. This provides a pure sample of the protein coded by the cDNA.

Hybrid-arrest translation is slightly different in that the denatured cDNA is added directly to the mRNA sample (Figure 11.21). Hybridization again occurs between the cDNA and its mRNA counterpart, but in this case the unbound mRNA is not discarded. Instead the entire sample is translated in the cell-free system. The hybridized mRNA is unable to direct translation, so all the proteins except the one coded by the cloned gene are synthesized. The cloned gene's translation product is therefore identified as the protein that is absent from the autoradiograph.

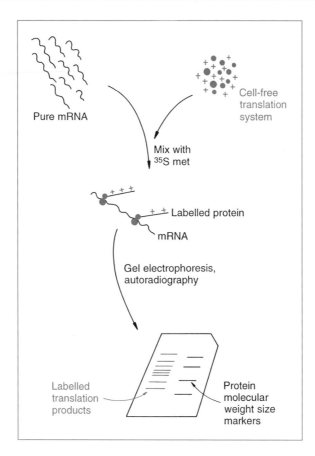

Figure 11.19 Cell-free translation.

Pure mRNA

Cell-free translation system

Mix with
^{35}S met

Labelled protein

mRNA

Gel electrophoresis,
autoradiography

Labelled translation products

Protein molecular weight size markers

11.3.2

Analysis of proteins by *in vitro* mutagenesis

Although HRT and HART can identify the translation product of a cloned gene, these techniques tell us little about the protein itself. The major questions asked by the molecular biologist today centre on the relationship between the structure of a protein and its mode of activity. The best way of tackling these problems is to induce a mutation in the gene coding for the protein and then to determine what effect the change in amino acid sequence has on the properties of the translation product (Figure 11.22). However, under normal circumstances mutations occur randomly and a large number may have to be screened before one that gives useful information is found. A solution to this problem is provided by *in vitro* **mutagenesis**, a technique that enables a directed mutation to be made at a specific point in a cloned gene.

Different types of *in vitro* mutagenesis techniques

An almost unlimited variety of DNA manipulations can be used to introduce mutations into cloned genes. The following are the simplest.

Figure 11.20 Hybrid-release translation.

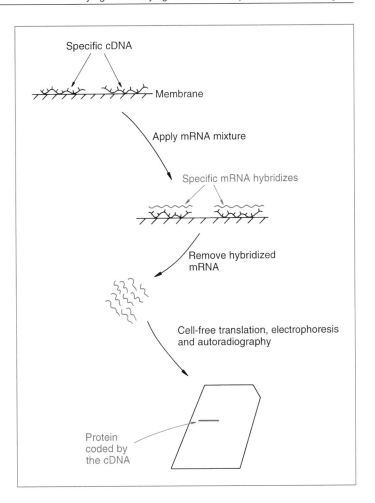

Specific cDNA

Membrane

Apply mRNA mixture

Specific mRNA hybridizes

Remove hybridized mRNA

Cell-free translation, electrophoresis and autoradiography

Protein coded by the cDNA

(1) A restriction fragment can be deleted (Figure 11.23(a)).

(2) The gene can be opened at a unique restriction site, a few nucleotides removed with a double-strand-specific endonuclease such as Bal31 (p. 53), and the gene religated (Figure 11.23(b)).

(3) A short oligonucleotide can be inserted at a restriction site (Figure 11.23(c)). The sequence of the oligonucleotide can be such that the additional stretch of amino acids inserted into the protein produces, for example, a new structure such as an α-helix, or destabilizes an existing structure.

Although potentially useful, these manipulations depend on the fortuitous occurrence of a restriction site at the area of interest in the cloned gene. **Oligonucleotide-directed mutagenesis** is a more versatile technique that can introduce a mutation at any point in the gene.

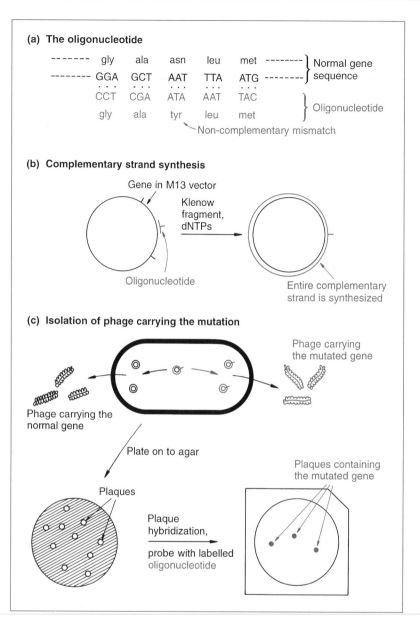

Figure 11.24 One method for oligonucleotide-directed mutagenesis.

The semi-conservative nature of DNA replication means that half the double-stranded molecules that are produced are unmutated in both strands, whereas half are mutated in both strands. Similarly, half the resulting phage progeny carry copies of the unmutated molecule and half carry the mutation. The phage produced by the transfected cells are plated onto solid agar so that plaques are produced. Half the plaques should contain the original recombinant molecule, and half the mutated version. Which are which is determined by plaque hybridization, using the oligonucleotide as the probe,

Figure 11.26 The yeast two hybrid system. (a) A pair of transcription factors that must interact in order for a yeast gene to be expressed. (b) Replacement of transcription factor 1 with the hybrid protein 1* abolishes gene expression as 1* cannot interact with transcription factor 2. (c) Replacement of transcription factor 2 with the hybrid protein 2* restores gene expression if the hybrid parts of 1* and 2* are able to interact.

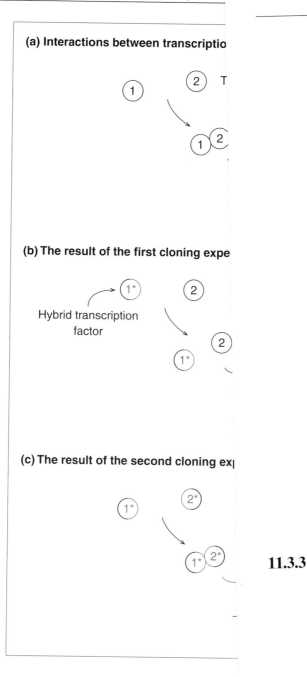

(a) Interactions between transcriptio

(b) The result of the first cloning expe

Hybrid transcription factor

(c) The result of the second cloning ex

libraries can be prepared by cloning a m
tissue or, less easily, by cloning genomic D
of phage displaying a range of different pr
that interact with a test protein. This test
one that is itself displayed on a phage sur
the wells of a microtitre tray or on partic
chromatography column, and then mixe

and employing very strict conditions so that only the completely base-paired hybrid is stable.

Cells infected with M13 vectors do not lyse, but instead continue to divide (p. 25). The mutated gene can therefore be expressed in the host *E. coli* cells resulting in production of recombinant protein. The protein coded by the mutated gene can be purified from the recombinant cells and its properties studied. The effect of a single base pair mutation on the activity of the protein can therefore be assessed.

The potential of oligonucleotide-directed mutagenesis

Oligonucleotide-directed mutagenesis and related techniques have remarkable potential, both for pure research and for applied biotechnology. For example, the biochemist can now ask very specific questions about the way that protein structure affects the action of an enzyme. In the past, it has been possible through biochemical analysis to gain some idea of the identity of the amino acids that provide the substrate binding and catalytic functions of an enzyme molecule. Mutagenesis techniques provide a much more detailed picture by enabling the role of each individual amino acid to be assessed by replacing it with an alternative residue and determining the effect this has on the enzymatic activity.

The ability to manipulate enzymes in this way has resulted in dramatic advances in our understanding of biological catalysis and has led to the new field of **protein engineering**, in which mutagenesis techniques are used to develop new enzymes for biotechnological purposes. For example, careful alterations to the amino acid sequence of subtilisin, an enzyme used in biological washing powders, have resulted in engineered versions with greater resistances to the thermal and bleaching (oxidative) stresses encountered in washing machines.

11.3.3

Studying protein–protein interactions

Within living cells, few if any proteins function in total isolation. Instead, proteins work together in biochemical pathways and in multiprotein complexes. Information about the function of a protein that has not previously been studied can often be obtained by determining which other proteins it works with in the cell. Two important techniques, **phage display** and the **yeast two hybrid system**, enable these protein–protein interactions to be examined.

Phage display

This technique is called phage display because it involves the 'display' of proteins on the surface of a bacteriophage, usually M13 (Figure 11.25(a)). This is achieved by cloning the gene for the protein in a special type of M13 vector, one that results in the cloned gene becoming fused with a gene for a phage

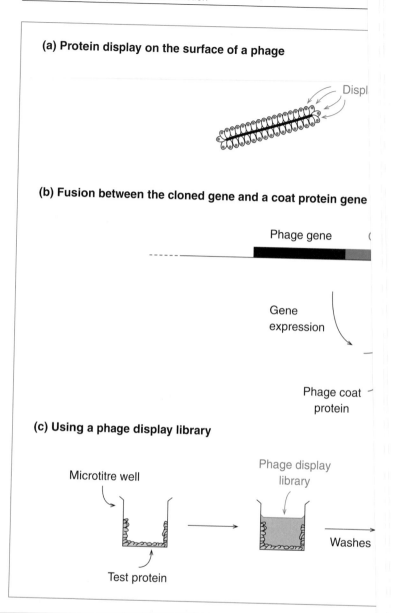

(a) Protein display on the surface of a phage

Displ

(b) Fusion between the cloned gene and a coat protein gene

Phage gene

Gene
expression

Phage coat
protein

(c) Using a phage display library

Microtitre well

Phage display
library

Washes

Test protein

Figure 11.25 Phage display. (a) Display of proteins on the surface of a
The gene fusion used to display a protein. (c) One way of detecting inte
phage from within a display library.

coat protein (Figure 11.25(b)). After tra
directs synthesis of a hybrid protein, ma
partly of the product of the cloned ger
be inserted into the phage coat so that t
located on the surface of the phage par
  Normally this technique is carried ou
up of many recombinant phage, each c

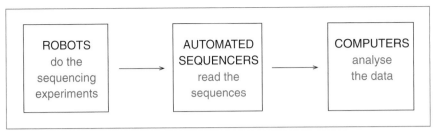

Figure 12.1 The factory
approach to large scale
DNA sequencing.

ROBOTS
do the
sequencing
experiments

AUTOMATED
SEQUENCERS
read the
sequences

COMPUTERS
analyse
the data

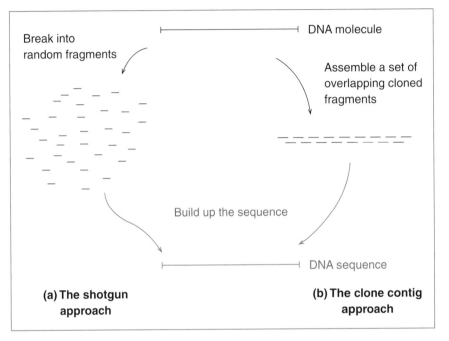

Figure 12.2 (a) The
shotgun and (b) the clone
contig approaches to
assembly of a genome
sequence.

Break into
random fragments

DNA molecule

Assemble a set of
overlapping cloned
fragments

Build up the sequence

DNA sequence

**(a) The shotgun
approach**

**(b) The clone contig
approach**

(2) The **clone contig approach**, which involves a presequencing phase during
which a series of overlapping clones is identified. Each piece of cloned
DNA is then sequenced, and this sequence placed at its appropriate position on the contig map in order to gradually build up the overlapping
genome sequence.

12.1.1

The shotgun approach to genome sequencing

The key requirement of the shotgun approach is that it must be possible to
identify overlaps between all the individual sequences that are generated,
and this identification process must be accurate and unambiguous so that the
correct genome sequence is obtained. An error in identifying a pair of overlapping sequences could lead to the genome sequence becoming scrambled,
or parts missed out entirely. The probability of making mistakes increases
with larger genome sizes, so the shotgun approach has been used mainly with
the smaller bacterial genomes.

Figure 12.3 A schematic of the key steps in the *H. influenzae* genome sequencing project.

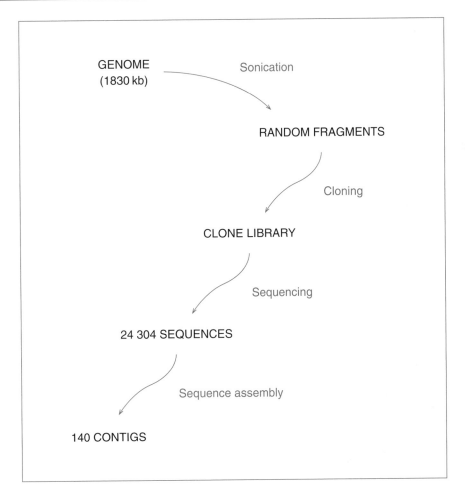

GENOME
(1830 kb)

Sonication

RANDOM FRAGMENTS

Cloning

CLONE LIBRARY

Sequencing

24 304 SEQUENCES

Sequence assembly

140 CONTIGS

The *H. influenzae* genome sequencing project

The shotgun approach was first used successfully with the bacterium *H. influenzae*, which was the first free-living organism whose genome was entirely sequenced, the results published in 1995. The first step was to break the 1830 kb genome of the bacterium into short fragments which would provide the templates for the sequencing experiments (Figure 12.3). A restriction endonuclease could have been used but sonication (p. 213) was chosen because this technique is more random and hence reduces the possibility of gaps appearing in the genome sequence.

It was decided to concentrate on fragments of 1.6–2.0 kb because these could yield two DNA sequences, one from each end, reducing the amount of cloning and DNA preparation that was required. The sonicated DNA was therefore fractionated by agarose gel electrophoresis and fragments of the desired size purified from the gel. After cloning, 28 643 sequencing experiments were carried out with 19 687 of the clones. A few of these sequences – 4339 in all – were rejected because they were less than 400 bp in length.

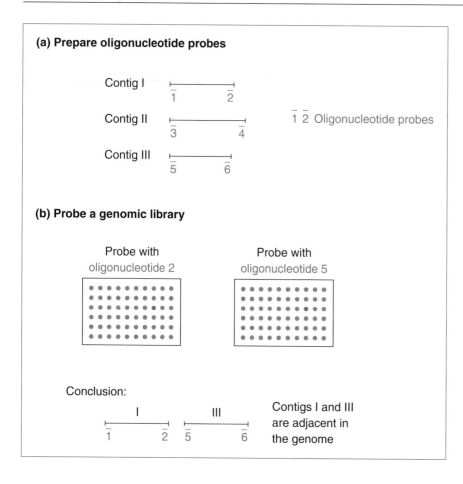

Figure 12.4 Using oligonucleotide hybridization to close gaps in the *H. influenzae* genome sequence. Oligonucleotides 2 and 5 both hybridize to the same λ clone, indicating that contigs I and III are adjacent. The gap between them can be closed by sequencing the appropriate part of the λ clone.

The remaining 24 304 sequences were entered into a computer which spent 30 hours analysing the data. The result was 140 contiguous sequences, each a different segment of the *H. influenzae* genome.

It might have been possible to continue sequencing more of the sonicated fragments in order eventually to close the gaps between the individual segments. However, 11 631 485 bp of sequence had already been generated – six times the length of the genome – suggesting that a large amount of additional work would be needed before the correct fragments were, by chance, sequenced. At this stage of the project the most time-effective approach was to use a more directed strategy in order to close each of the gaps individually. Several approaches were used for gap closure, the most successful of these involving hybridization analysis of a clone library prepared in a λ vector (Figure 12.4). The library was probed in turn with a series of oligonucleotides whose sequences corresponded with the ends of each of the 140 segments. In some cases, two oligonucleotides hybridized to the same λ clone, indicating that the two segment ends represented by those oligonucleotides lay adjacent to one another in the genome. The gap between these two ends could then be closed by sequencing the appropriate part of the λ clone.

Figure 12.5 One problem with the shotgun approach. An incorrect overlap is made between two sequences that both terminate within a repeated element. The result is that a segment of the DNA molecule is absent from the DNA sequence.

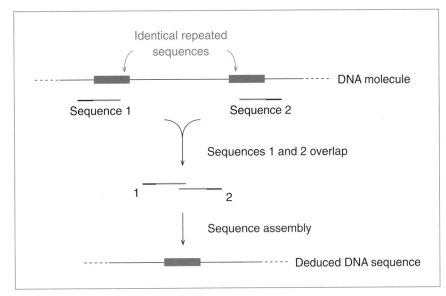

Problems with shotgun cloning

Shotgun cloning has been successful with many bacterial genomes. Not only are these genomes small, so the computational requirements for finding sequence overlaps are not too great, but they contain little or no repetitive DNA sequences. These are sequences, from a few base pairs to several kilobases, that are repeated at two or more places in a genome. They cause problems for the shotgun approach because when sequences are assembled those that lie partly or wholly within one repeat element might accidentally be assigned an overlap with the identical sequence present in a different repeat element (Figure 12.5). This could lead to parts of the genome sequence being assembled in the incorrect position or left out entirely. For this reason, it has generally been thought that shotgun sequencing is inappropriate for eukaryotic genomes, as these have many repeat elements. Later in the chapter we will see how this limitation can be circumvented using a genome map to direct assembly of sequences obtained by the shotgun approach.

12.1.2

The clone contig approach

The clone contig approach does not suffer from the limitations of shotgun sequencing and so can provide an accurate sequence of a large genome that contains repetitive DNA. Its drawback is that it involves much more work and so takes longer and costs more money. The additional time and effort is needed to construct the overlapping series of cloned DNA fragments. Once this has been done, each cloned fragment is sequenced by the shotgun method and the genome sequence built up step by step (Figure 12.2).

The cloned fragments should be as long as possible in order to minimize the total number needed to cover the entire genome. A high capacity vector is therefore used. The first eukaryotic chromosome to be sequenced – chromosome III of *Saccharomyces cerevisiae* – was initially cloned in a cosmid vector (p. 125) with the resulting contig comprising 29 clones. Chromosome III is relatively short, however, and the average size of the cloned fragments was just 10.8 kb. Sequencing of the much longer human genome required 300 000 bacterial artificial chromosome (BAC) clones (p. 127). Assembling all of these into chromosome-specific contigs was a massive task.

Clone contig assembly by chromosome walking

One technique that can be used to assemble a clone contig is **chromosome walking**. To begin a chromosome walk a clone is selected at random from the library, labelled, and used as a hybridization probe against all the other clones in the library (Figure 12.6(a)). Those clones that give hybridization signals are ones that overlap with the probe. One of these overlapping clones is now labelled and a second round of probing carried out. More hybridization signals are seen, some of these indicating additional overlaps (Figure 12.6(b)). Gradually the clone contig is built up in a step-by-step fashion. But this is a laborious process and is only attempted when the contig is for a short chromosome and so involves relatively few clones, or when the aim is to close one or more small gaps between contigs that have been built up by more rapid methods.

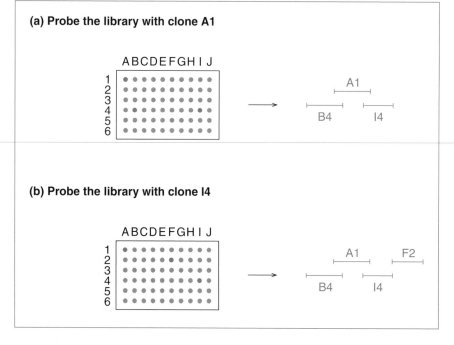

Figure 12.6 Chromosome walking.

Rapid methods for clone contig assembly

The weakness of chromosome walking is that it begins at a fixed starting point and builds up the clone contig step by step, and hence slowly, from that fixed point. The more rapid techniques for clone contig assembly do not use a fixed starting point and instead aim to identify pairs of overlapping clones: when enough overlapping pairs have been identified the contig is revealed (Figure 12.7). The various techniques that can be used to identify overlaps are collectively known as **clone fingerprinting**.

Clone fingerprinting is based on the identification of sequence features that are shared by a pair of clones. The simplest approach is to digest each clone with one or more restriction endonucleases and to look for pairs of clones that share restriction fragments of the same size, excluding those fragments that derive from the vector rather than the inserted DNA. This technique might appear to be easy to carry out but in practice it takes a great deal of time to scan the resulting agarose gels for shared fragments. There is also a relatively high possibility that two clones that do not overlap will, by chance, share restriction fragments whose sizes are indistinguishable by agarose gel electrophoresis.

More accurate results can be obtained by **repetitive DNA PCR**, also know as **interspersed repeat element PCR (IRE–PCR)**. This type of PCR uses primers that are designed to anneal within repetitive DNA sequences and direct amplification of the DNA between adjacent repeats (Figure 12.8). Repeats of a particular type are distributed fairly randomly in eukaryotic genomes, with varying distances between them, so a variety of product sizes are obtained when these primers are used with clones of eukaryotic DNA. If a pair of clones give PCR products of the same size, they must contain repeats that are identically spaced, possibly because the cloned DNA fragments overlap.

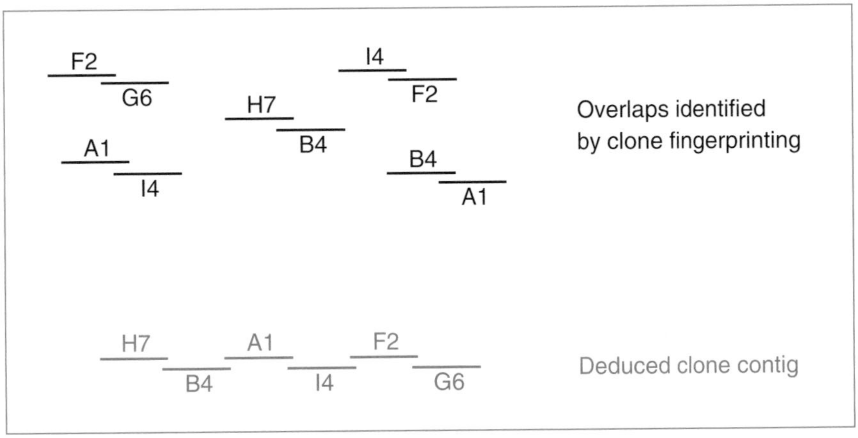

Figure 12.7 Building up a clone contig by a clone fingerprinting technique.

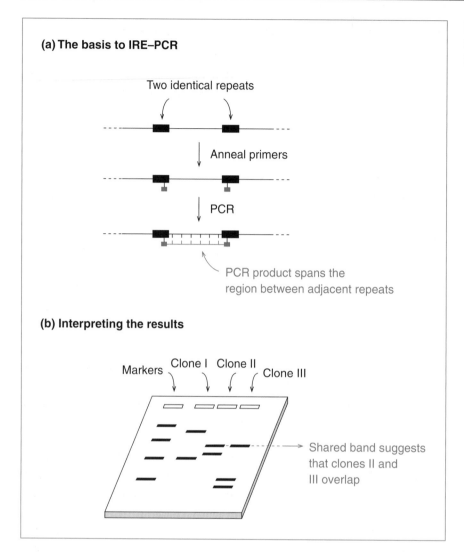

(a) The basis to IRE–PCR

Two identical repeats

Anneal primers

PCR

PCR product spans the
region between adjacent repeats

(b) Interpreting the results

Markers   Clone I   Clone II   Clone III

Shared band suggests
that clones II and
III overlap

Figure 12.8 Interspersed repeat element PCR (IRE–PCR).

Clone contig assembly by sequence tagged site content analysis

A third way to assemble a clone contig is to search for pairs of clones that contain a specific DNA sequence that occurs at just one position in the genome under study. If two clones contain this feature then clearly they must overlap (Figure 12.9). A sequence of this type is called a **sequence tagged site (STS)**. Often an STS is a gene that has been sequenced in an earlier project. As the sequence is known, a pair of PCR primers can be designed that are specific for that gene and then used to identify which members of a clone library contain the gene. But the STS does not have to be a gene and can be any short piece of DNA sequence that has been obtained from the genome, providing it does not fall in a repetitive element.

12.1.3

Using a map to aid sequence assembly

Sequence tagged site content mapping is a particularly important method for clone contig assembly because often the positions of STSs within the genome will have been determined by **genetic mapping** or **physical mapping**. This means that the STS positions can be used to anchor the clone contig onto a genome map, enabling the position of the contig within a chromosome to be determined. We will now look at how these maps are obtained.

Genetic maps

A genetic map is one that is obtained by genetic studies using Mendelian principles and involving directed breeding programmes for experimental organisms or **pedigree analysis** for humans. In most cases the loci that are studied are genes, whose inheritance patterns are followed by monitoring the phenotypes of the offspring produced after a cross between parents with contrasting characteristics (e.g. tall and short for the pea plants studied by Mendel). More recently, techniques have been devised for genetic mapping of DNA sequences that are not genes but which display variability in the human population. The most important of these **DNA markers** are:

(1) **Restriction fragment length polymorphisms (RFLPs)**, which are caused by a sequence variation that results in a restriction site being changed. When digested with a restriction endonuclease the loss of the site is revealed because two fragments remain joined together. Originally, RFLPs were typed by Southern hybridization of restricted genomic DNA, but this is a time-consuming process, so nowadays the presence or absence of the restriction site is usually determined by PCR (Figure 12.10). There are approximately 100 000 RFLPs in the human genome.

(2) **Short tandem repeats (STRs)**, also called **microsatellites**, which are made up of short repetitive sequences, such as CACACA. The repeat units are 1–13 nucleotides in length and are usually repeated 5–20 times. The number of repeats at a locus can be determined by carrying out a PCR using primers that anneal either side of the STR, and examining

Figure 12.9 The basis to STS content mapping.

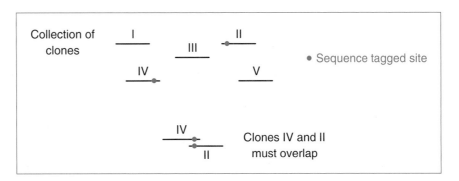

255

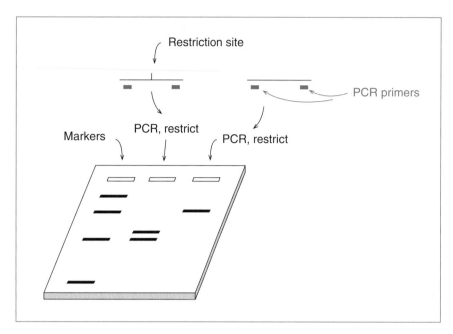

Figure 12.10 Typing a restriction site polymorphism by PCR. In the middle lane the PCR product gives two bands because it is cut by treatment with the restriction enzyme. In the right-hand lane there is just one band because the template DNA lacks the restriction site.

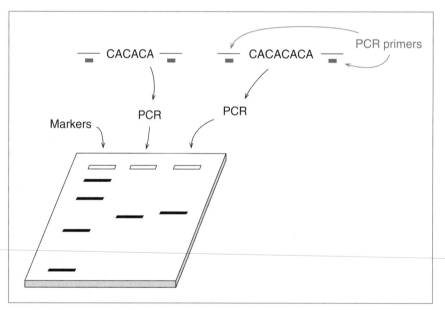

Figure 12.11 Typing an STR by PCR. The PCR product in the right-hand lane is slightly longer than that in the middle lane, because the template DNA from which it is generated contains an additional CA unit.

the size of the resulting products by agarose or polyacrylamide gel electrophoresis (Figure 12.11). There are at least 650 000 STRs in the human genome.

(3) Single nucleotide polymorphisms (SNPs), which are positions in a genome where any one of two or more different nucleotides can occur (Figure 12.12). These point mutations are typed by analysis with short

Figure 12.12 Two versions of an SNP.

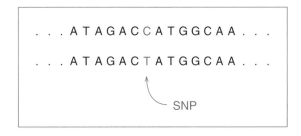

... A T A G A C C A T G G C A A ...

... A T A G A C T A T G G C A A ...

SNP

oligonucleotide probes that hybridize to the alternative forms of the SNP. The number of SNPs in the human genome is not yet known but is at least 1.4 million.

All of these DNA markers are variable and so exist in two or more allelic forms. The inheritance of the alternative alleles at a particular locus is followed by analysis of DNA prepared from the offspring of genetic crosses.

Physical maps

A physical map is generated by methods that directly locate the positions of specific sequences on a chromosomal DNA molecule. As in genetic mapping, the loci that are studied can be genes or DNA markers. The latter might include **expressed sequence tags (ESTs)**, which are short sequences obtained from the ends of complementary DNAs (cDNAs) (p. 164). Expressed sequence tags are therefore partial gene sequences, and when used in map construction they provide a quick way of locating the positions of genes, even though the identity of the gene might not be apparent from the EST sequence. Two types of technique are used in physical mapping.

(1) Direct examination of chromosomal DNA molecules, for example by fluorescence *in situ* hybridization (FISH) (p. 203). If FISH is carried out simultaneously with two DNA probes, each labelled with a different fluorochrome, the relevant positions on the chromosome of the two markers represented by the probe can be visualized. Special techniques for working with extended chromosomes, whose DNA molecules are stretched out rather than tightly coiled as in normal chromosomes, enable markers to be positioned with a high degree of accuracy.

(2) Physical mapping with a **mapping reagent**, which is a collection of overlapping DNA fragments spanning the chromosome or genome that is being studied. Pairs of markers that lie within a single fragment must be located close to each other on the chromosome: how close can be determined by measuring the frequency with which the pair occurs together in different fragments in the mapping reagent

Figure 12.13 The principle behind the use of a mapping reagent. It can be deduced that markers 1 and 2 are relatively close because they are present together on four DNA fragments. In contrast, markers 3 and 4 must be relatively far apart because they occur together on just one fragment.

(Figure 12.13). **Radiation hybrids** are one type of mapping reagent that has been important in the Human Genome Project. These are hamster cell lines that contain fragments of human chromosomes, prepared by a treatment involving irradiation (hence their name). Mapping is carried out by hybridization of marker probes to a panel of cell lines, each one containing a different part of the human genome.

The importance of a map in sequence assembly

It is possible to obtain a genome sequence without the use of a genetic or physical map. This is illustrated by the *H. influenzae* project that we followed on p. 249, and many other bacterial genomes have been sequenced without the aid of a map. But a map is very important when a larger genome is being sequenced because it provides a guide that can be used to check that the genome sequence is being assembled correctly from the many short sequences that emerge from the automated sequencer. If a marker that has been mapped by genetic and/or physical means appears in the genome sequence at an unexpected position then an error in sequence assembly is suspected.

Detailed genetic and/or physical maps have been important in the Human Genome Project, as well as those for yeast, fruit fly, *C. elegans* and *A. thaliana*, all of which were based on the clone contig approach. Maps are also being used to direct sequence assembly in projects that use the shotgun approach. As described on p. 251, the major problem when applying shotgun sequencing to a large genome is the presence of repeated sequences and the possibility that the assembled sequence 'jumps' between two repeats, so part of the genome is misplaced or left out (Figure 12.5). These errors can be avoided if sequence assembly makes constant reference to a genome map. This 'directed shotgun approach' holds great promise as a rapid method of sequencing large genomes.

12.2

Post-genomics – trying to understand a genome sequence

Once a genome sequence has been completed the next step is to locate all the genes and determine their functions. This is a far from trivial process, even with genomes that have been extensively studied by genetic analysis and gene cloning techniques prior to complete sequencing. For example, the sequence of *S. cerevisiae*, one of the best studied of all organisms, revealed that this genome contains about 6000 genes. Of these, some 3600 could be assigned a function either on the basis of previous studies that had been carried out with yeast or because the yeast gene had a similar sequence to a gene that had been studied in another organism. This left 2400 genes whose functions were not known.

Despite a massive amount of work since the yeast genome was completed in 1996, the functions of the vast majority of these **orphans** have still not been determined. It is in this area that bioinformatics, sometimes referred to as molecular biology *in silico*, is proving of major value as an adjunct to conventional experiments.

12.2.1

Identifying the genes in a genome sequence

Locating a gene is easy if the amino acid sequence of the protein product is known, allowing the nucleotide sequence of the gene to be predicted, or if the corresponding cDNA or EST has been previously sequenced. But for many genes there is no prior information that enables the correct DNA sequence to be recognized. Under these circumstances, gene location might be difficult even if a map is available. Most maps have only a limited accuracy and can only delineate the approximate position of a gene, possibly leaving several tens or even hundreds of kilobases to search in order to find it. And many genes do not appear on maps because their existence is unsuspected. How can these genes be located in a genome sequence?

Searching for open reading frames

The DNA sequence of a gene is an **open reading frame (ORF)**, a series of nucleotide triplets beginning with an initiation codon (usually but not always AUG) and ending in a termination codon (TAA, TAG or TGA in most genomes). Searching a genome sequence for ORFs, by eye or more usually by computer, is therefore the first step in gene location. It is important to search all six **reading frames** because genes can run in either direction along the DNA double helix (Figure 12.14(a)). With a bacterial genome the typical result of this search is identification of long ORFs that are almost certainly genes, with many shorter ORFs partly or completely contained within the genes but lying in different reading frames (Figure 12.14(b)). These short

(a) Every DNA sequence has six reading frames

```
1        G A C →
2      T G A →
3    A T G →
5' – A T G A C C A A T G A C A T G C A A T A A – 3'
      | | | | | | | | | | | | | | | | | | | | | | |
3' – T A C T G G T T A C T G T A C G T T A T T – 5'
                              ← A T T   4
                            ← T A T     5
                          ← T T A       6
```

(b) The typical result of an ORF search with a bacterial genome

Genes
Spurious ORFs

100 bp

Figure 12.14 Searching for open reading frames. (a) Every DNA sequence has six open reading frames, any one of which could contain a gene. (b) The typical result of a search for ORFs in a bacterial genome. The arrows indicate the directions in which the genes and spurious ORFs run.

sequences are almost certainly combinations of nucleotides that by chance form an ORF but are not genes. If one of these short ORFs lies entirely between two genes there is a possibility that it might mistakenly be identified as a real gene, but in most bacterial genomes there is very little space between the genes so the problem arises only infrequently.

Gene location in eukaryotes is much more difficult. Eukaryotic genomes are not as densely packed as bacterial ones and there are much longer spacers between genes. This means that inspection of the sequence reveals many short ORFs that cannot be discounted because they do not overlap with real genes. Analysis of the yeast genome, for example, identified over 400 short ORFs that were placed in this 'questionable' category. Possibly some of these are real genes but probably most of them are not.

In humans and other higher eukaryotes, the search for genes is made even more complicated by the fact that many are split into exons and introns (Figure 11.1). Particular nucleotide sequences always occur at exon–intron boundaries (Figure 12.15) but these sequences are also found within exons and within introns. Working out which of these sequences mark true exon–intron boundaries can be very difficult.

Distinguishing real genes from chance ORFS

Some genomes provide helpful signposts that indicate the presence of a gene. The human and other vertebrate genomes are particularly helpful to the molecular biologist because 50–60% of the genes are accompanied by a **CpG**

Figure 12.15 The consensus sequences for the upstream and downstream exon–intron boundaries of vertebrate introns. Py = pyrimidine nucleotide (C or T), N = any nucleotide. The arrows are the boundary positions.

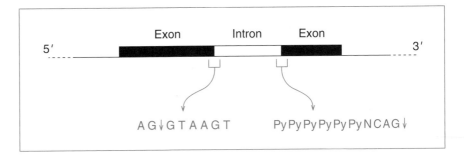

island, a distinctive GC-rich sequence whose position indicates the approximate start point for a gene. But features such as these are the exception rather than the rule and more general methods for identifying genes are needed.

With many genomes, **codon bias** provides a useful means of assigning a degree of certainty to a possible gene identification. All amino acids except methionine and tryptophan are specified by two or more codons. Alanine, for example, has four codons – GCA, GCC, GCG and GCT. In most genomes, not all members of a codon family are used with equal frequency. Humans are typical in this regard, displaying a distinct bias for certain codons: for example, within the alanine codon family, humans use GCC four times more frequently than GCG. If an ORF contains a high frequency of rare codons then it probably is not a gene. By taking account of the codon bias displayed by an ORF an informed guess can therefore be made as to whether the sequence is or is not a gene.

Tentative identification of a gene is usually followed by a **homology search**. This is an analysis, carried out by computer, in which the sequence of the gene is compared with all the gene sequences present in the international DNA databases, not just known genes of the organism under study but also genes from all other species. The rationale is that two genes from different organisms that have similar functions have similar sequences, reflecting their common evolutionary histories (Figure 12.16).

To carry out a homology search the nucleotide sequence of the tentative gene is usually translated into an amino acid sequence, as this allows a more sensitive search. This is because there are 20 different amino acids but only four nucleotides, so there is less chance of two amino acid sequences appearing to be similar purely by chance.

The analysis is carried out through the internet, by logging on to the web site of one of the DNA databases and using a search programme such as **BLAST** (Basic Local Alignment Search Tool). If the test sequence is over 200 amino acids in length and has 30% or greater identity with a sequence in the database (i.e. at 30 out of 100 positions the same amino acid occurs in both sequences), then the two are almost certainly homologous and the ORF under study can be confirmed as a real gene. Further confirmation, if needed,

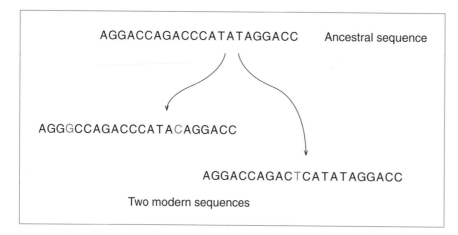

AGGACCAGACCCATATAGGACC   Ancestral sequence

AGGGCCAGACCCATACAGGACC

AGGACCAGACTCATATAGGACC

Two modern sequences

Figure 12.16 Homology between two sequences that share a common ancestor. The two sequences have acquired mutations during their evolutionary histories but their sequence similarities indicate that they are homologues.

can be obtained by using transcript analysis (p. 218) to show that the gene is transcribed into (RNA).

12.2.2 Determining the function of an unknown gene

Homology searching serves two purposes. As well as testing the veracity of a tentative gene identification it can also give an indication of the function of the gene, presuming that the function of the homologous gene is known. Almost 2000 of the genes in the yeast genome were assigned functions in this way. Frequently, however, the matches found by homology searching are to other genes whose functions have yet to be determined. These unassigned genes are called orphans and working out their function is one of the major challenges of post-genomics research.

In future years it will probably be possible to use bioinformatics to gain at least an insight into the function of an orphan gene. It is already possible to use the nucleotide sequence of a gene to predict the positions of α-helices and β-sheets in the encoded protein, albeit with limited accuracy, and the resulting structural information can sometimes be used to make inferences about the function of the protein. Proteins that attach to membranes can often be identified because they possess α-helical arrangements that span the membrane, and DNA binding motifs such as zinc fingers can be recognized. A greater scope and accuracy to this aspect of bioinformatics will be possible when more information is obtained about the relationship between the structure of a protein and its function. In the meantime, functional analysis of orphans depends largely on conventional experiments.

Several techniques for studying the function of genes were described in Chapter 11, and all of these can be applied to orphans. One strategy not described in Chapter 11 is **gene knockout**. In this technique a deleted version of the gene is used to 'knock out' the functional version present in the organism's chromosomes. This is possible because recombination between the deleted gene, carried on a cloning vector, and the chromosomal copy can

Figure 12.17 Gene knockout by recombination between a chromosomal copy of a gene and a deleted version carried by a plasmid cloning vector.

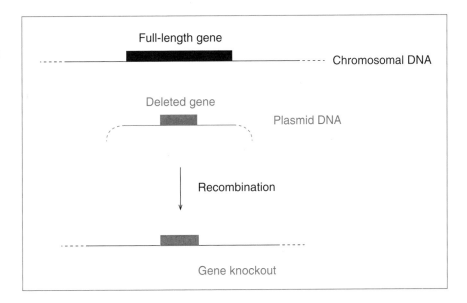

result in the former replacing the latter (Figure 12.17). The effect of the gene knockout on the phenotype of the organism is then assessed in order to gain some insight into the function of the gene. The effect of a human gene knockout is inferred by studying a **knockout mouse**, which has a deleted version of the homologous mouse gene. Knockouts have helped to assign functions to a number of genes but problems can arise. In particular, some knockouts have no obvious effect on the phenotype of the organism either because the gene is dispensable, meaning that after inactivation other genes can compensate for its absence, or because the phenotypic change is too subtle to be detected.

12.3 Studies of the transcriptome and proteome

So far we have considered those aspects of post-genomics research that are concerned with studies of individual genes. The change in emphasis from genes to the genome has prompted new types of analysis that are aimed at understanding the activity of the genome as a whole. This work has led to the invention of two new terms:

(1) The **transcriptome**, which is the messenger RNA (mRNA) content of a cell, and which reflects the overall pattern of gene expression in that cell.

(2) The **proteome**, which is the protein content of a cell and which reflects its biochemical capability.

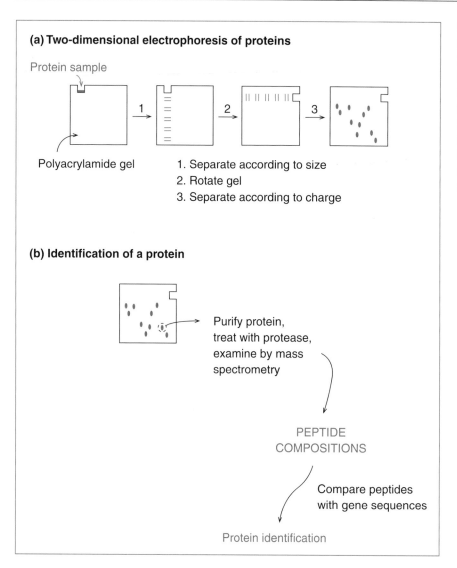

(a) Two-dimensional electrophoresis of proteins

Protein sample

Polyacrylamide gel

1. Separate according to size
2. Rotate gel
3. Separate according to charge

(b) Identification of a protein

Purify protein,
treat with protease,
examine by mass
spectrometry

PEPTIDE
COMPOSITIONS

Compare peptides
with gene sequences

Protein identification

Figure 12.21 Proteome analysis. (a) Two-dimensional polyacrylamide gel electrophoresis of proteins. (b) Identification of the protein contained in a single spot by protease treatment followed by mass spectrometry of the resulting peptides.

would then reveal which genes are up- or down-regulated in response to the cancerous state.

An alternative to microarrays is provided by **DNA chips**, which are thin wafers of silicon that carry many different oligonucleotides (Figure 12.19). These oligonucleotides are synthesized directly on the surface of the chip and can be prepared at a density of 1 million per cm^2, substantially higher than is possible with a conventional microarray. Hybridization between an oligonucleotide and the probe is detected electronically. Because the oligonucleotides are synthesized *de novo*, using special automated procedures, a chip carrying oligonucleotides whose sequences are specific for every human gene is relatively easy to prepare.

12.3.2 Studying the proteome

The proteome is the entire collection of proteins in a cell. Proteome studies provide additional information that is not obtainable simply by examining the transcriptome, because a single mRNA (and hence gene) can give rise to more than one protein, because of post-translational processing (Figure 12.20, p. 265). In eukaryotes, most of the polypeptides that are synthesized by translation are further processed by addition of chemical groups. The particular additions that are made determine the precise function of the protein. Phosphorylation, for example, is an important modification used to activate some proteins.

To study the proteome the entire protein content of a cell or tissue is first separated by two-dimensional electrophoresis. In this technique, the proteins are loaded into a well on one side of a square of polyacrylamide gel and separated according to their molecular weights. The square is then rotated by 90° and a second electrophoresis performed, this time separating the proteins on the basis of their charges. The result is a two-dimensional pattern of spots, of different sizes, shapes and intensities, each representing a different protein or related group of proteins (Figure 12.21(a)). Differences between two proteomes are apparent from differences in the pattern of spots when the two gels are compared. To identify the protein in a particular spot a sample is purified from the gel and treated with a protease that cuts the polypeptide at a specific amino acid sequence (similar in a way to the activity of a restriction endonuclease). The resulting peptides are then examined by mass spectrometry (Figure 12.21(b)). The mass spectrometer determines the amino acid composition of each peptide. This information is usually sufficient to enable the gene coding for the protein to be identified from the genome sequence.

Further reading

Adams, M.D., Celnicker, S.E., Holt, R.A. *et al.* (2000) The genome sequence of *Drosophila melanogaster*. *Science*, **287**, 2185–95. [A clear description of this genome project.]

Brown, T.A. (1999) *Genomes*. BIOS Scientific Publishers, Oxford. [Gives details of techniques for studying genomes, including genetic and physical mapping.]

Deloukas, P., Schuler, G.D., Gyapay, G. *et al.* (1998) A physical map of 30 000 genes. *Science*, **282**, 744–6. [A map of the human genome.]

DeRisi, J.L., Iyer, V.R. & Brown, P.O. (1997) Exploring the metabolic and genetic control of gene expression on a genomic scale. *Science*, **278**, 680–86. [One of the first uses of microarray analysis to study a transcriptome.]

Dujon, B. (1996) The yeast genome project: what did we learn? *Trends in Genetics*, **12**, 263–70. [Describes the analysis of the yeast genome sequence and some of the problems in assigning functions to genes.]

Fleischmann, R.D., Adams, M.D., White, O. *et al.* (1995) Whole genome random sequencing and assembly of *Haemophilus influenzae* Rd. *Science*, **269**, 496–512. [The first complete bacterial genome sequence to be published.]

van Hal, N.L.W., Vorst, O., van Houwelingen, A.M.M.L. *et al.* (2000) The application of DNA microarrays in gene expression analysis. *Journal of Biotechnology*, **78**, 271–80.

International Human Genome Sequencing Consortium (2001) Initial sequencing and analysis of the human genome. *Nature*, **409**, 860–921.

Li, L. (2000) Progress in proteomics. *Progress in Biochemistry and Biophysics*, **27**, 227–31.

Ramsay, G. (1998) DNA chips: state of the art. *Nature Biotechnology*, **16**, 40–4.

Smith, T.F. (1998) Functional genomics – bioinformatics is ready for the challenge. *Trends in Genetics*, **14**, 291–3. [A summary of computer tools for studying genomes.]

The *Arabidopsis* Genome Initiative (2000) Analysis of the genome sequence of the flowering plant *Arabidopsis thaliana*. *Nature*, **408**, 796–815.

Walter, M.A., Spillett, D.J., Thomas, P., Weissenbach, J. & Goodfellow, P.N. (1994) A method for constructing radiation hybrid maps of whole genomes. *Nature Genetics*, **7**, 22–8.

Wilson, D.R. & Finlay, B.B. (1998) Phage display: applications, innovations, and issues in phage and host biology. *Canadian Journal of Microbiology*, **44**, 313–29.

PART 3
THE APPLICATIONS OF GENE CLONING AND DNA ANALYSIS IN BIOTECHNOLOGY

Chapter 13 **Production of Protein from Cloned Genes**

Now that we have covered the basic techniques involved in gene cloning and DNA analysis and examined how these techniques are used in research, we can move on to consider how recombinant DNA technology is being applied in biotechnology. This is not a new subject, although biotechnology has received far more attention during the last few years than it ever has in the past. Biotechnology can be defined as the use of biological processes in industry and technology. According to archaeologists, the British biotechnology industry dates back 4000 years, to the late Neolithic period, when fermentation processes that make use of living yeast cells to produce ale and mead were first introduced into this country. Certainly brewing was well established by the time of the Roman invasion.

During the twentieth century, biotechnology expanded with the development of a variety of industrial uses for microorganisms. The discovery by Alexander Fleming in 1929 that the mould *Penicillium* synthesizes a potent antibacterial agent led to the use of fungi and bacteria in the large scale production of antibiotics. At first the microorganisms were grown in large culture vessels from which the antibiotic was purified after the cells had been removed (Figure 13.1(a)), but more recently this **batch culture** method has been largely supplanted by **continuous culture** techniques, making use of a **fermenter**, from which samples of medium can be continuously drawn off, providing a non-stop supply of the product (Figure 13.1(b)). This type of process is not limited to antibiotic production and has also been used to obtain large amounts of other compounds produced by microorganisms (Table 13.1).

One of the reasons why biotechnology has received so much attention during the last decade is because of gene cloning. Although many useful products can be obtained from microbial culture, the list in the past has been limited to those compounds naturally synthesized by microorganisms. Many important pharmaceuticals that are produced not by microbes but by higher

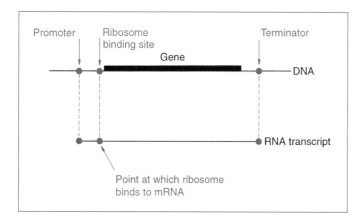

Figure 13.3 The three most important signals for gene expression in *E. coli*.

recombinant protein will be synthesized. This is because expression is dependent on the gene being surrounded by a collection of signals that can be recognized by the bacterium. These signals, which are short sequences of nucleotides, advertise the presence of the gene and provide instructions for the transcriptional and translational apparatus of the cell. The three most important signals for *E. coli* genes are as follows (Figure 13.3):

(1) The **promoter**, which marks the point at which transcription of the gene should start. In *E. coli* the promoter is recognized by the σ subunit of the transcribing enzyme RNA polymerase.

(2) The **terminator**, which marks the point at the end of the gene where transcription should stop. A terminator is usually a nucleotide sequence that can base pair with itself to form a **stem loop** structure.

(3) The **ribosome binding site**, a short nucleotide sequence recognized by the ribosome as the point at which it should attach to the messenger RNA (mRNA) molecule. The initiation codon of the gene is always a few nucleotides downstream of this site.

The genes of higher organisms are also surrounded by expression signals, but their nucleotide sequences are not the same as the *E. coli* versions. This is illustrated by comparing the promoters of *E. coli* and human genes (Figure 13.4). There are similarities, but it is unlikely that an *E. coli* RNA polymerase would be able to attach to a human promoter. A foreign gene is inactive in *E. coli* quite simply because the bacterium does not recognize its expression signals.

A solution to this problem would be to insert the foreign gene into the vector in such a way that it is placed under control of a set of *E. coli* expression signals. If this can be achieved the gene should be transcribed and translated (Figure 13.5). Cloning vehicles that provide these signals, and can therefore be used in the production of recombinant protein, are called **expression vectors**.

Figure 13.4 Typical promoter sequences for *E. coli* and animal genes.

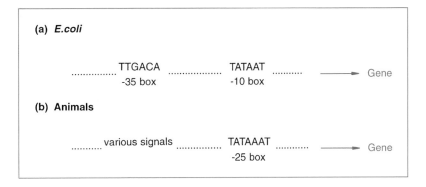

(a) **E.coli**

TTGACA .... TATAAT .... ⟶ Gene
-35 box .... -10 box

(b) **Animals**

.... various signals .... TATAAAT .... ⟶ Gene
-25 box

Figure 13.5 The use of an expression vector to achieve expression of a foreign gene in *E. coli*.

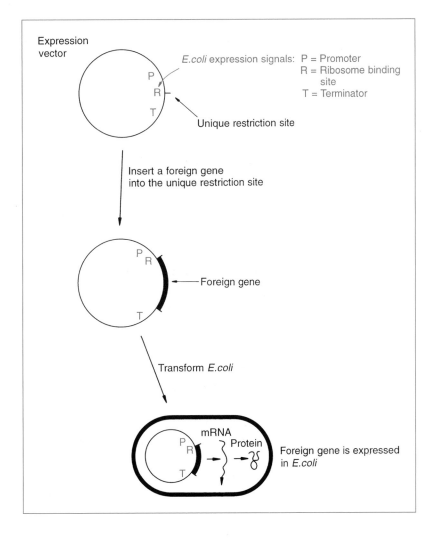

Expression vector

E.coli expression signals: P = Promoter
R = Ribosome binding site
T = Terminator

P
R
T

Unique restriction site

Insert a foreign gene into the unique restriction site

P R
T

Foreign gene

Transform *E.coli*

mRNA
P R   Protein
T

Foreign gene is expressed in *E.coli*

13.1.1

The promoter is the critical component of an expression vector

The promoter is the most important component of an expression vector. This is because the promoter controls the very first stage of gene expression (attachment of an RNA polymerase enzyme to the DNA) and determines the rate at which mRNA is synthesized. The amount of recombinant protein obtained therefore depends to a great extent on the nature of the promoter carried by the expression vector.

The promoter must be chosen with care

The two sequences shown in Figure 13.4(a) are **consensus sequences**, averages of all the *E. coli* promoter sequences that are known. Although most *E. coli* promoters do not differ much from these consensus sequences (e.g. TTTACA instead of TTGACA), a small variation may have a major effect on the efficiency with which the promoter can direct transcription. **Strong promoters** are those that can sustain a high rate of transcription; strong promoters usually control genes whose translation products are required in large amounts by the cell (Figure 13.6(a)). In contrast, **weak promoters**, which are relatively inefficient, direct transcription of genes whose products are needed in only small amounts (Figure 13.6(b)). Clearly an expression vector should carry a strong promoter, so that the cloned gene is transcribed at the highest possible rate.

A second factor to be considered when constructing an expression vector is whether it will be possible to regulate the promoter in any way. Two major types of gene regulation are recognized in *E. coli* – **induction** and **repression**. An inducible gene is one whose transcription is switched on by addition of a chemical to the growth medium; often this chemical is one of the substrates for the enzyme coded by the inducible gene (Figure 13.7(a)). In contrast, a repressible gene is switched off by addition of the regulatory chemical (Figure 13.7(b)).

Gene regulation is a complex process that only indirectly involves the promoter itself. However, many of the sequences important in induction and repression lie in the region surrounding the promoter and are therefore also present in an expression vector. It may therefore be possible to extend the regulation to the expression vector, so that the chemical that induces or represses the gene normally controlled by the promoter is also able to regulate expression of the cloned gene. This can be a distinct advantage in the production of recombinant protein. For example, if the recombinant protein has a harmful effect on the bacterium, then its synthesis must be carefully monitored to prevent accumulation of toxic levels: this can be achieved by judicious use of the regulatory chemical to control expression of the cloned gene. Even if the recombinant protein has no harmful effects on the host cell, regulation of the cloned gene is still desirable, as a continuously high level

Figure 13.6 Strong and weak promoters.

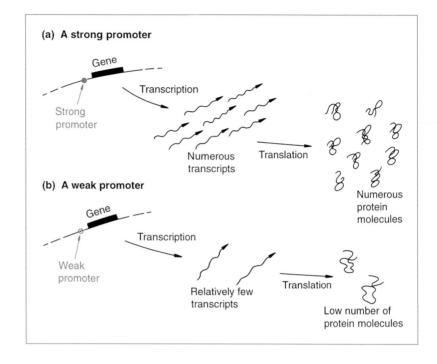

(a) **A strong promoter**

Gene

Strong promoter

Transcription

Numerous transcripts

Translation

Numerous protein molecules

(b) **A weak promoter**

Gene

Weak promoter

Transcription

Relatively few transcripts

Translation

Low number of protein molecules

Figure 13.7 Examples of the two major types of gene regulation that occur in bacteria: (a) an inducible gene, and (b) a repressible gene.

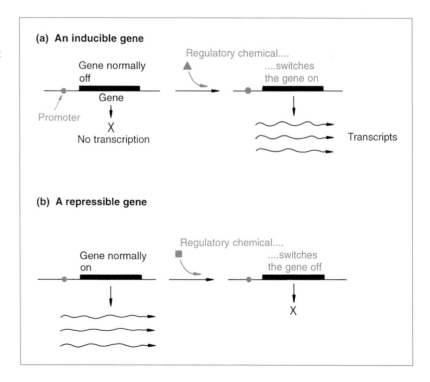

(a) **An inducible gene**

Gene normally off

Promoter

Gene

X

No transcription

Regulatory chemical....

....switches the gene on

Transcripts

(b) **A repressible gene**

Gene normally on

Regulatory chemical....

....switches the gene off

X

of transcription may affect the ability of the recombinant plasmid to replicate, leading to its eventual loss from the culture.

Examples of promoters used in expression vectors

Several *E. coli* promoters combine the desired features of strength and ease of regulation. Those most frequently used in expression vectors are as follows.

(1) The **lac promoter** (Figure 13.8(a)) is the sequence that controls transcription of the *lacZ* gene coding for β-galactosidase (and also the *lacZ'* gene fragment carried by the pUC and M13mp vectors; pp. 108 and 112). The *lac* promoter is induced by isopropylthiogalactoside (IPTG, p. 95), so addition of this chemical into the growth medium switches on transcription of a gene inserted downstream of the *lac* promoter carried by an expression vector.

(2) The **trp promoter** (Figure 13.8(b)) is normally upstream of the cluster of genes coding for several of the enzymes involved in biosynthesis of the amino acid tryptophan. The *trp* promoter is repressed by tryptophan, but is more easily induced by 3-β-indoleacrylic acid.

(3) The **tac promoter** (Figure 13.8(c)) is a hybrid between the *trp* and *lac* promoters. It is stronger than either, but still induced by IPTG.

(4) The **λP$_L$ promoter** (Figure 13.8(d)) is one of the promoters responsible for transcription of the λ DNA molecule. λP$_L$ is a very strong promoter that is recognized by the *E. coli* RNA polymerase, which is subverted

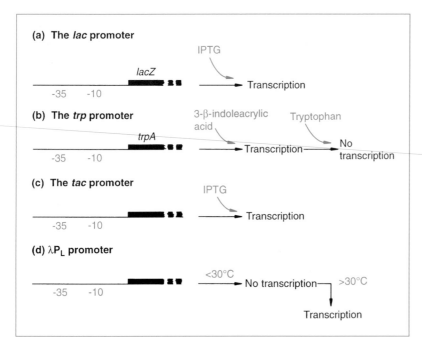

(a) **The *lac* promoter**

IPTG

lacZ

-35   -10

Transcription

(b) **The *trp* promoter**

3-β-indoleacrylic acid

Tryptophan

trpA

-35   -10

Transcription ——— No transcription

(c) **The *tac* promoter**

IPTG

-35   -10

Transcription

(d) λP$_L$ **promoter**

<30°C

-35   -10

No transcription ——— >30°C

Transcription

Figure 13.8 Four promoters frequently used in expression vectors. The *lac* and *trp* promoters are shown upstream of the genes that they normally control in *E. coli*.

by λ into transcribing the bacteriophage DNA. The promoter is repressed by the product of the λ*c*I gene. Expression vectors that carry the λP$_L$ promoter are used with a mutant *E. coli* host that synthesizes a temperature-sensitive form of the *c*I protein (p. 46). At a low temperature (less than 30°C) this mutant *c*I protein is able to repress the λP$_L$ promoter; at higher temperatures the protein is inactivated, resulting in transcription of the cloned gene.

13.1.2

Cassettes and gene fusions

An efficient expression vector requires not only a strong, regulatable promoter, but also an *E. coli* ribosome binding sequence and a terminator. In most vectors these expression signals form a **cassette**, so-called because the foreign gene is inserted into a unique restriction site present in the middle of the expression signal cluster (Figure 13.9). Ligation of the foreign gene into the cassette therefore places it in the ideal position relative to the expression signals.

With some cassette vectors the cloning site is not immediately adjacent to the ribosome binding sequence, but instead is preceded by a segment from the beginning of an *E. coli* gene (Figure 13.10). Insertion of the foreign gene into this restriction site must be performed in such a way as to fuse the two reading frames, producing a hybrid gene that starts with the *E. coli* segment and progresses without a break into the codons of the foreign gene. The product of gene expression is therefore a hybrid protein, consisting of the short peptide coded by the *E. coli* reading frame fused to the amino-terminus of the foreign protein. This fusion system has four advantages:

(1) Efficient translation of the mRNA produced from the cloned gene depends not only on the presence of a ribosome binding site, but is also affected by the nucleotide sequence at the start of the coding region. This is probably because secondary structures resulting from intrastrand base pairs could interfere with attachment of the ribosome to its binding site (Figure 13.11). This possibility is avoided if the pertinent region is made up entirely of natural *E. coli* sequences.

Figure 13.9 A typical cassette vector and the way it is used. P = promoter, R = ribosome binding site, T = terminator.

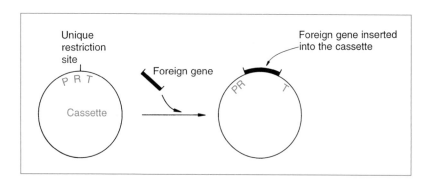

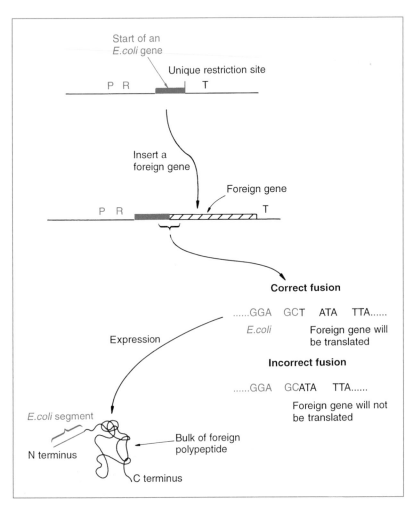

Figure 13.10 The construction of a hybrid gene and the synthesis of a fusion protein.

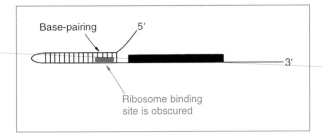

Figure 13.11 A problem caused by secondary structure at the start of an mRNA.

(2) The presence of the bacterial peptide at the start of the fusion protein may stabilize the molecule and prevent it from being degraded by the host cell. In contrast, foreign proteins that lack a bacterial segment are often destroyed by the host.

(3) The bacterial segment may constitute a signal peptide, responsible for directing the *E. coli* protein to its correct position in the cell. If the signal

Figure 13.12 The use of affinity chromatography to purify a glutathione-*S*-transferase fusion protein.

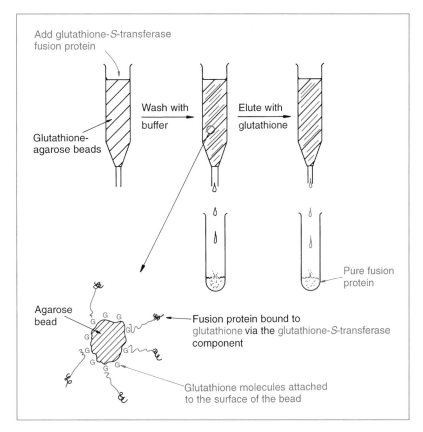

peptide is derived from a protein that is exported by the cell (e.g. the products of the *ompA* or *malE* genes), the recombinant protein may itself be exported, either into the culture medium or into the periplasmic space between the inner and outer cell membranes. Export is desirable as it simplifies the problem of purification of the recombinant protein from the culture.

(4) The bacterial segment may also aid purification by enabling the fusion protein to be recovered by **affinity chromatography**. For example, fusions involving the *E. coli* glutathione-*S*-transferase protein can be purified by adsorption onto agarose beads carrying bound glutathione (Figure 13.12).

The disadvantage with a fusion system is that the presence of the *E. coli* segment may alter the properties of the recombinant protein. Methods for removing the bacterial segment are therefore needed. Usually this is achieved by treating the fusion protein with a chemical or enzyme that cleaves the polypeptide chain at or near the junction between the two components. For example, if a methionine is present at the junction, the fusion protein can be cleaved with cyanogen bromide, which cuts polypeptides specifically at methionine residues (Figure 13.13). Alternatively, enzymes

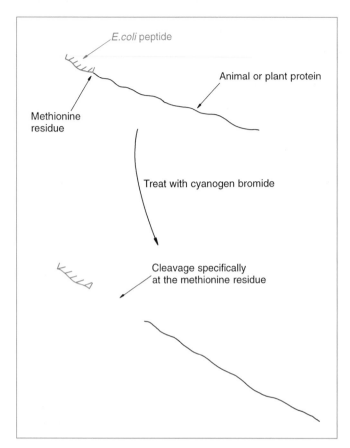

E.coli peptide

Animal or plant protein

Methionine
residue

Treat with cyanogen bromide

Cleavage specifically
at the methionine residue

Figure 13.13 One method for the recovery of the foreign polypeptide from a fusion protein. The methionine residue at the fusion junction must be the only one present in the entire polypeptide: if others are present cyanogen bromide will cleave the fusion protein into more than two fragments.

such as thrombin (which cleaves adjacent to arginine residues) or factor Xa (which cuts after the arginine of Gly–Arg) can be used. The important consideration is that recognition sequences for the cleavage agent must not occur within the recombinant protein.

13.2 General problems with the production of recombinant protein in *E. coli*

Despite the development of sophisticated expression vectors, there are still numerous difficulties associated with the production of protein from foreign genes cloned in *E. coli*. These problems can be grouped into two categories: those that are due to the sequence of the foreign gene, and those that are due to the limitations of *E. coli* as a host for recombinant protein synthesis.

13.2.1   Problems resulting from the sequence of the foreign gene

There are three ways in which the nucleotide sequence might prevent efficient expression of a foreign gene cloned in *E. coli*:

(1)   The foreign gene might contain introns. This would be a major problem as *E. coli* genes do not contain introns and therefore the bacterium does not possess the necessary machinery for removing introns from transcripts (Figure 13.14(a)).

(2)   The foreign gene might contain sequences that act as termination signals in *E. coli* (Figure 13.14(b)). These sequences are perfectly innocuous in the normal host cell but in the bacterium result in premature termination and a loss of gene expression.

(3)   The codon bias of the gene may not be ideal for translation in *E. coli*. As described on p. 261, although virtually all organisms use the same genetic code, each organism has a bias towards preferred codons. This bias reflects the efficiency with which the transfer RNA (tRNA) molecules in the organism are able to recognize the different codons. If a cloned gene contains a high proportion of unfavoured codons, the host cell's tRNAs may encounter difficulties in translating the gene, reducing the amount of protein that is synthesized (Figure 13.14(c)).

These problems can usually be solved, although the necessary manipulations may be time-consuming and costly (an important consideration in an industrial project). If the gene contains introns then its complementary DNA (cDNA), prepared from the mRNA (p. 164) and so lacking introns, might be used as an alternative. Oligonucleotide-directed mutagenesis could then be employed to change the sequences of possible terminators and to replace unfavoured codons with those preferred by *E. coli*. An alternative with genes that are less than 2 kb in length is to make an artificial version. This involves synthesizing a set of overlapping oligonucleotides (p. 172) that are ligated together, the sequences of the oligonucleotides being designed to ensure that the resulting gene contains preferred *E. coli* codons and that terminators are absent.

13.2.2   Problems caused by *E. coli*

Some of the difficulties encountered when using *E. coli* as the host for recombinant protein synthesis stem from inherent properties of the bacterium. For example:

(1)   *E. coli* might not process the recombinant protein correctly. The proteins of most organisms are processed after translation, by chemical modification of amino acids within the polypeptide. Often these processing events are essential for the correct biological activity of the protein. Unfortunately, the proteins of bacteria and higher organisms

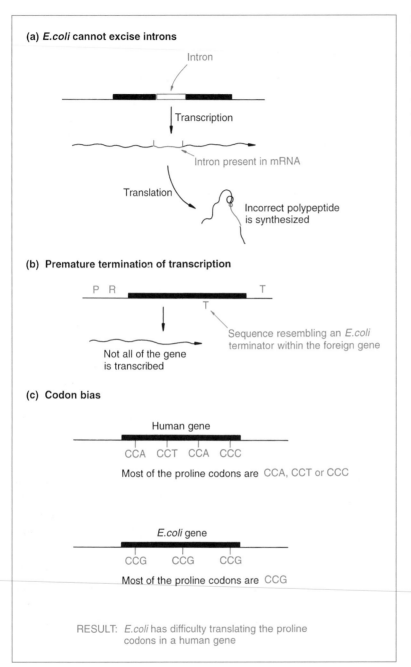

(a) *E.coli* cannot excise introns

Intron

Transcription

Intron present in mRNA

Translation

Incorrect polypeptide
is synthesized

(b) Premature termination of transcription

P R   T

T

Sequence resembling an *E.coli*
terminator within the foreign gene

Not all of the gene
is transcribed

(c) Codon bias

Human gene

CCA CCT CCA CCC

Most of the proline codons are CCA, CCT or CCC

E.coli gene

CCG   CCG   CCG

Most of the proline codons are CCG

RESULT: *E.coli* has difficulty translating the proline
codons in a human gene

Figure 13.14 Three of the problems that could be encountered when foreign genes are expressed in *E. coli*. (a) Introns are not removed in *E. coli*. (b) Premature termination of transcription. (c) A problem with codon bias.

are not processed identically. In particular, some animal proteins are glycosylated, meaning that they have sugar groups attached to them after translation. Glycosylation is extremely uncommon in bacteria and recombinant proteins synthesized in *E. coli* are never glycosylated correctly.

(2) *E. coli* might not fold the recombinant protein correctly, and generally

Figure 13.15 Inclusion bodies.

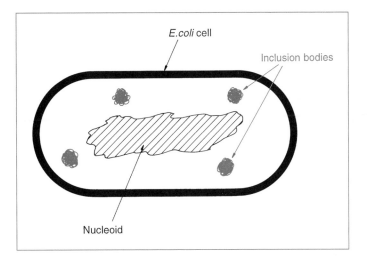

is unable to synthesize the disulphide bonds present in many animal proteins. If the protein does not take up its correctly folded tertiary structure then usually it is insoluble and forms an **inclusion body** within the bacterium (Figure 13.15). Recovery of the protein from the inclusion body is not a problem, but converting the protein into its correctly folded form is difficult or impossible in the test tube. Under these circumstances the protein is, of course, inactive.

(3) *E. coli* might degrade the recombinant protein. Exactly how *E. coli* can recognize the foreign protein, and thereby subject it to preferential turnover, is not known.

These problems are less easy to solve than the sequence problems described in the previous section. Degradation of recombinant proteins can be reduced by using as the host a mutant *E. coli* strain that is deficient in one or more of the proteases responsible for protein degradation. Correct folding of recombinant proteins can also be promoted by choosing a special host strain, in this case one that oversynthesizes the chaperone proteins thought to be responsible for protein folding in the cell. But the main problem is the absence of glycosylation. So far this has proved insurmountable, limiting *E. coli* to the synthesis of animal proteins that do not need to be processed in this way.

13.3 Production of recombinant protein by eukaryotic cells

The problems associated with obtaining high yields of active recombinant proteins from genes cloned in *E. coli* have led to the development of expression systems for other organisms. There have been a few attempts to use other bacteria as the hosts for recombinant protein synthesis, and some progress has been made with *Bacillus subtilis*, but the main alternatives to *E. coli* are microbial eukaryotes. The argument is that a microbial eukaryote, such as a yeast or filamentous fungus, is more closely related to an animal, and so may be able to deal with recombinant protein synthesis more efficiently than *E. coli* can. Yeasts and fungi can be grown just as easily as bacteria in continuous culture, and might express a cloned gene from a higher organism, and process the resulting protein, in a manner more akin to that occurring in the higher organism itself.

13.3.1 Recombinant protein from yeast and filamentous fungi

To a large extent the potential of microbial eukaryotes has been realized and these organisms are now being used for the routine production of several animal proteins. Expression vectors are still required because it turns out that the promoters and other expression signals for animal genes do not in general work efficiently in these lower eukaryotes. The vectors themselves are based on those described in Chapter 7.

Saccharomyces cerevisiae as the host for recombinant protein synthesis

The yeast *Saccharomyces cerevisiae* is currently the most popular microbial eukaryote for recombinant protein production. Cloned genes are often placed under the control of the *GAL* promoter (Figure 13.16(a)), which is normally upstream of the gene coding for galactose epimerase, an enzyme involved in the metabolism of galactose. The *GAL* promoter is induced by galactose, providing a straightforward system for regulating expression of a cloned foreign gene. Other useful promoters are *PHO5*, which is regulated by the phosphate level in the growth medium, and *CUP1*, which is induced by copper. Most yeast expression vectors also carry a termination sequence from an *S. cerevisiae* gene because animal termination signals do not work effectively in yeast.

Yields of recombinant protein are relatively high, but *S. cerevisiae* is unable to glycosylate animal proteins correctly, often adding too many sugar units ('hyperglycosylation'), although this can be prevented or at least reduced by using a mutant host strain. *S. cerevisiae* also lacks an efficient system for secreting proteins into the growth medium. In the absence of secretion, recombinant proteins are retained in the cell and consequently are less easy to purify. Codon bias (p. 283) can also be a problem.

Despite these drawbacks, *S. cerevisiae* remains the most frequently used microbial eukaryote for recombinant protein synthesis, partly because it is accepted as a safe organism for production of proteins for use in medicines or in foods, and partly because of the wealth of knowledge built up over the years regarding the biochemistry and genetics of *S. cerevisiae*, which means

Figure 13.16 Four promoters frequently used in expression vectors for microbial eukaryotes. P = promoter.

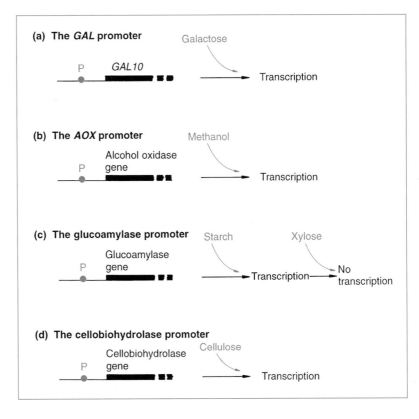

(a) **The *GAL* promoter**

Galactose

P   GAL10

Transcription

(b) **The *AOX* promoter**

Methanol

Alcohol oxidase
P   gene

Transcription

(c) **The glucoamylase promoter**

Starch   Xylose

Glucoamylase
P   gene

Transcription — No transcription

(d) **The cellobiohydrolase promoter**

Cellulose

Cellobiohydrolase
P   gene

Transcription

that it is relatively easy to devise strategies for minimizing the difficulties that arise.

Other yeasts and fungi

Although *S. cerevisiae* retains the loyalty of many molecular biologists, there are other microbial eukaryotes that might be equally if not more effective in recombinant protein synthesis. In particular, *Pichia pastoris*, a second species of yeast, is able to synthesize large amounts of recombinant protein (up to 30% of the total cell protein) and its glycosylation abilities are very similar to those of animal cells. The sugar structures that it synthesizes are not precisely the same as the animal versions (Figure 13.17), but the differences are relatively trivial and would probably not have a significant effect on the activity of a recombinant protein. Importantly, the glycosylated proteins made by *P. pastoris* are unlikely to induce an antigenic reaction if injected into the bloodstream, a problem frequently encountered with the overglycosylated proteins synthesized by *S. cerevisiae*. Expression vectors for *P. pastoris* make use of the alcohol oxidase (*AOX*) promoter (Figure 13.16(b)), which is induced by methanol. The only significant problem with *P. pastoris* is that it sometimes degrades recombinant proteins before they can be purified, but this can be controlled by using special growth media. Other yeasts that have

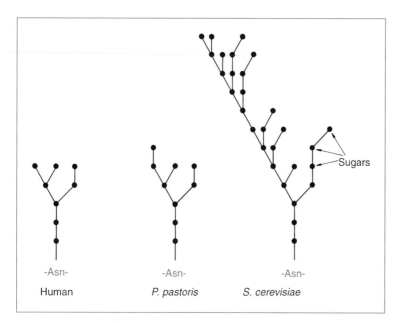

Figure 13.17 Comparison between a typical glycosylation structure found on an animal protein and the structures synthesized by *P. pastoris* and *S. cerevisiae*.

-Asn-

Human

-Asn-

P. pastoris

-Asn-

S. cerevisiae

Sugars

been used for recombinant protein synthesis include *Hansenula polymorpha*, *Yarrowia lipolytica* and *Kluveromyces lactis*. The last of these has the attraction that it can be grown on waste products from the food industry.

The two most popular filamentous fungi are *Aspergillus nidulans* and *Trichoderma reesei*. The advantages of these organisms are their good glycosylation properties and their ability to secrete proteins into the growth medium. The latter is a particularly strong feature of the wood rot fungus *T. reesei*, which in its natural habitat secretes cellulolytic enzymes that degrade the wood it lives on. The secretion characteristics mean that these fungi are able to produce recombinant proteins in a form that aids purification. Expression vectors for *A. nidulans* usually carry the glucoamylase promoter (Figure 13.16(c)), induced by starch and repressed by xylose; those for *T. reesei* make use of the cellobiohydrolase promoter (Figure 13.16(d)), which is induced by cellulose.

13.3.2   Using animal cells for recombinant protein production

The difficulties inherent in synthesis of a fully active animal protein in a microbial host have prompted biotechnologists to explore the possibility of using animal cells for recombinant protein synthesis. For proteins with complex and essential glycosylation structures, an animal cell might be the only type of host within which the active protein can be synthesized.

Protein production in mammalian cells

Culture systems for animal cells have been around since the early 1960s, but only during the last ten years have methods for large scale continuous culture become available. A problem with some animal cell lines is that they require a solid surface on which to grow, adding complications to the design of the culture vessels. One solution is to fill the inside of the vessel with plates, providing a large surface area, but this has the disadvantage that complete and continuous mixing of the medium within the vessel becomes very difficult. A second possibility is to use a standard vessel but to provide the cells with small inert particles (e.g. cellulose beads) on which to grow. Rates of growth and maximum cell densities are much less for animal cells compared with microorganisms, limiting the yield of recombinant protein, but this can be tolerated if it is the only way of obtaining the active protein.

Of course, gene cloning may not be necessary in order to obtain an animal protein from an animal cell culture. Nevertheless, expression vectors and cloned genes are still used to maximize yields, by placing the gene under control of a promoter that is stronger than the one it is normally attached to. This promoter is often obtained from a virus such as SV40 (p. 152). Mammalian cell lines derived from humans or hamsters have been used in synthesis of several recombinant proteins, and in all cases these proteins have been processed correctly and are indistinguishable from the non-recombinant versions. However, this is the most expensive approach to recombinant protein production, especially as the possible copurification of viruses with the protein means that rigorous quality control procedures must be employed to ensure that the product is safe.

Protein production in insect cells

Insect cells provide an alternative to mammalian cells for animal protein production. Insect cells do not behave in culture any differently to mammalian cells but they have the great advantage that, thanks to a natural expression system, they can provide high yields of recombinant protein.

The expression system is based on the **baculoviruses**, a group of viruses that are common in insects but apparently do not infect vertebrates. The baculovirus genome includes the polyhedrin gene, whose normal product accumulates in the insect cell as large nuclear inclusion bodies towards the end of the infection cycle (Figure 13.18). The product of this single gene frequently makes up over 50% of the total cell protein. Similar levels of protein production also occur if the normal gene is replaced by a foreign one. Baculovirus vectors have been successfully used in production of a number of mammalian proteins, but unfortunately the resulting proteins are not glycosylated correctly. In this regard the baculovirus system does not offer any advantages compared with *S. cerevisiae* or *P. pastoris*.

Pharming – recombinant protein from live animals

A recent and potentially important innovation in recombinant protein production is to use a **transgenic animal**, an animal that contains a cloned gene in all of its cells, usually introduced by microinjection of a fertilized egg cell

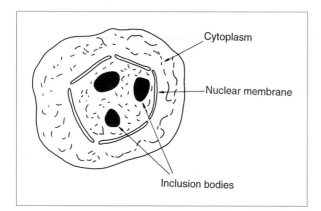

Figure 13.18 Crystalline inclusion bodies in the nuclei of insect cells infected with a baculovirus.

(p. 153). Transgenic animals are expensive to produce but the technology is cost-effective, because once a transgenic animal has been made it can reproduce and pass its cloned gene to its offspring according to standard Mendelian principles.

The most successful approach to recombinant protein production in transgenic animals has been to use farm animals such as sheep or pigs, with the cloned gene attached to the promoter for the animal's β-lactoglobulin gene. This promoter is active in the mammary tissue which means that the recombinant protein is secreted in the milk (Figure 13.19). Milk production can be continuous during the animal's adult life, resulting in a high yield of the protein, and because the protein is secreted, purification is relatively easy. Most importantly, sheep and pigs are mammals and so human proteins produced in this way are correctly modified. Production of pharmaceutical proteins in farm animals has, not surprisingly, been called **pharming**. It is a controversial technique but one that offers considerable promise for synthesis of correctly modified human proteins for use in medicine.

13.3.3

Recombinant proteins from plants

Plants provide the final possibility for production of recombinant protein. Plants and animals have similar protein processing activities, and most animal proteins produced in plants undergo the correct post-translation modifications and so are fully functional. Plant cell culture is a well established technology that is already used in the commercial synthesis of natural plant products. Alternatively, intact plants can be grown to a high density in fields, giving good yields of the recombinant protein and the potential for long-term storage in organs such as tubers or fruits which are naturally rich in protein.

Whichever production system is used, plants offer a cheap and low-technology means of mass production of recombinant proteins. A range of proteins has been produced in experimental systems, including important pharmaceuticals such as interleukins and antibodies. This is an area of intensive research at the present time, with a number of plant biotechnology

Figure 13.19
Recombinant protein production in the milk of a transgenic sheep.

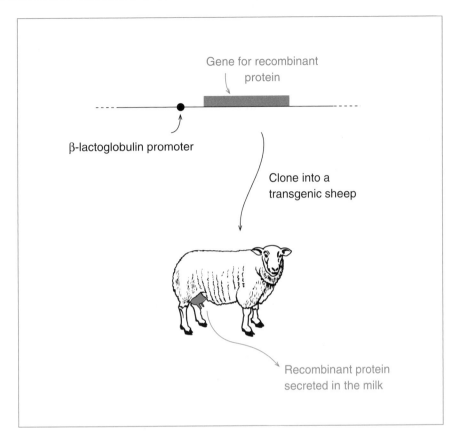

Gene for recombinant protein

β-lactoglobulin promoter

Clone into a transgenic sheep

Recombinant protein secreted in the milk

companies developing systems that are nearing commercial production. One very promising future possibility is that plants could be used to synthesize vaccines, providing the basis to a cheap and efficient vaccination programme.

Further reading

Baneyx, F. (1999) Recombinant protein expression in *Escherichia coli. Current Opinion in Biotechnology*, **10**, 411–21. [A general review of the use of *E. coli* in this area.]

de Boer, H.A., Comstock, L.J. & Vasser, M. (1983) The *tac* promoter: a functional hybrid derived from the *trp* and *lac* promoters. *Proceedings of the National Academy of Sciences of the USA*, **80**, 21–5.

Fischer, R. & Emans, N. (2000) Molecular farming of pharmaceutical proteins. *Transgenic Research*, **9**, 279–99. [A review of recombinant protein production in plants.]

Gellissen, G. & Hollenberg, C.P. (1997) Applications of yeasts in gene expression studies: a comparison of *Saccharomyces cerevisiae, Hansenula polymorpha* and *Kluveromyces lactis* – a review. *Gene*, **190**, 87–97.

Giddings, G., Allison, G., Brooks, D. & Carter, A. (2000) Transgenic plants as factories for biopharmaceuticals. *Nature Biotechnology*, **18**, 1151–5.

Hannig, G. & Makrides, S.C. (1998) Strategies for optimizing heterologous protein expression in *Escherichia coli. Trends in Biotechnology*, **16**, 54–60.

Kost, T.A. & Condreay, J.P. (1999) Recombinant baculoviruses as expression vectors for insect and mammalian cells. *Current Opinion in Biotechnology*, **10**, 428–33.

Lee, N., Cozzikorto, J., Wainwright, N. & Testa, D. (1984) Cloning with tandem gene systems for high level gene expression. *Nucleic Acids Research*, **12**, 6797–812. [Fusion proteins.]

Lubon, H. & Palmer, C. (2000) Transgenic animal bioreactors – where we are. *Transgenic Research*, **9**, 301–4. [An overview of pharming.]

Meerman, H.J. & Georgiou, G. (1994) Construction and characterization of a set of *E. coli* strains deficient in all known loci affecting the proteolytic stability of secreted recombinant proteins. *Biotechnology*, **12**, 1107–10. [Countering the problems caused by degradation of recombinant proteins in *E. coli*.]

Remaut, E., Stanssens, P. & Fiers, W. (1981) Plasmid vectors for high-efficiency expression controlled by the P_L promoter of coliphage. *Gene*, **15**, 81–93. [Construction of an expression vector.]

Robinson, M., Lilley, R., Little, S. *et al.* (1984) Codon usage can affect efficiency of translation of genes in *Escherichia coli. Nucleic Acids Research*, **12**, 6663–71.

Smith, D.B. & Johnson, K.S. (1988) Single step purification of polypeptides expressed in *Escherichia coli* as fusions with glutathione *S*-transferase. *Gene*, **67**, 3110.

Sreekrishna, K., Brankamp, R.G., Kropp, K.E. *et al.* (1997) Strategies for optimal synthesis and secretion of heterologous proteins in the methylotrophic yeast *Pichia pastoris*. *Gene*, **190**, 55–62.

Wurm, F. & Bernard, A. (1999) Large scale transient expression in mammalian cells for recombinant protein production. *Current Opinion in Biotechnology*, **10**, 156–9.

Chapter 14 Gene Cloning and DNA Analysis in Medicine

Medicine has been and will continue to be a major beneficiary of the recombinant DNA revolution, and an entire book could be written on this topic. Later in this chapter we will see how recombinant DNA techniques are being used to identify genes responsible for inherited diseases and to devise new therapies for these disorders. First we will continue the theme developed in the last chapter and examine the ways in which cloned genes are being used in the production of recombinant pharmaceuticals.

14.1 Production of recombinant pharmaceuticals

A number of human disorders can be traced to the absence or malfunction of a protein normally synthesized in the body. Most of these disorders can be treated by supplying the patient with the correct version of the protein, but for this to be possible the relevant protein must be available in relatively large amounts. If the defect can be corrected only by administering the human protein, then obtaining sufficient quantities will be a major problem unless donated blood can be used as the source. Animal proteins are therefore used whenever these are effective, but there are not many disorders that can be treated with animal proteins, and there is always the possibility of side effects such as an allergenic response.

We learned in Chapter 13 that gene cloning can be used to obtain large amounts of recombinant human proteins. How are these techniques being applied to the production of proteins for use as pharmaceuticals?

14.1.1 Recombinant insulin

Insulin, synthesized by the β-cells of the islets of Langerhans in the pancreas, controls the level of glucose in the blood. An insulin deficiency manifests itself as diabetes mellitus, a complex of symptoms which may lead to death if untreated. Fortunately, many forms of diabetes can be alleviated by a continuing programme of insulin injections, thereby supplementing the limited amount of hormone synthesized by the patient's pancreas. The insulin used in this treatment has traditionally been obtained from the pancreas of pigs and cows slaughtered for meat production. Although animal insulin is generally satisfactory, problems may arise in its use to treat human diabetes. One problem is that the slight differences between the animal and the human proteins can lead to side effects in some patients. Another is that the purification procedures are difficult, and potentially dangerous contaminants cannot always be completely removed.

Insulin displays two features that facilitate its production by recombinant DNA techniques. The first is that the human protein is not modified after translation by the addition of sugar molecules (p. 284); recombinant insulin synthesized by a bacterium should therefore be active. The second advantage concerns the size of the molecule. Insulin is a relatively small protein, comprising two polypeptides, one of 21 amino acids (the A chain) and one of 30 amino acids (the B chain; Figure 14.1). In humans these chains are synthesized as a precursor called preproinsulin, which contains the A and B segments linked by a third chain (C) and preceded by a leader sequence. The leader sequence is removed after translation and the C chain excised, leaving the A and B polypeptides linked to each other by two disulphide bonds.

Several strategies have been used to obtain recombinant insulin. One of the first projects, involving synthesis of artificial genes for the A and B chains followed by production of fusion proteins in *E. coli*, illustrates a number of the general techniques used in recombinant protein production.

Synthesis and expression of artificial insulin genes

In the late 1970s the idea of making an artificial gene was extremely innovative. Oligonucleotide synthesis was in its infancy at that time, and the available methods for making artificial DNA molecules were much more cumbersome than the present day automated techniques. Nevertheless, genes coding for the A and B chains of insulin were synthesized as early as 1978.

The procedure used was to synthesize trinucleotides representing all the possible codons and then join these together in the order dictated by the amino acid sequences of the A and B chains. The artificial genes would not necessarily have the same nucleotide sequences as the real gene segments coding for the A and B chains, but they would still specify the correct polypeptides. Two recombinant plasmids were constructed, one carrying the artificial gene for the A chain and one the gene for the B chain.

In each case the artificial gene was ligated to a *lacZ'* reading frame present in a pBR322-type vector (Figure 14.2(a)). The insulin genes were therefore under the control of the strong *lac* promoter (p. 278), and were

Figure 14.1 The structure of the insulin molecule and a summary of its synthesis by processing from preproinsulin.

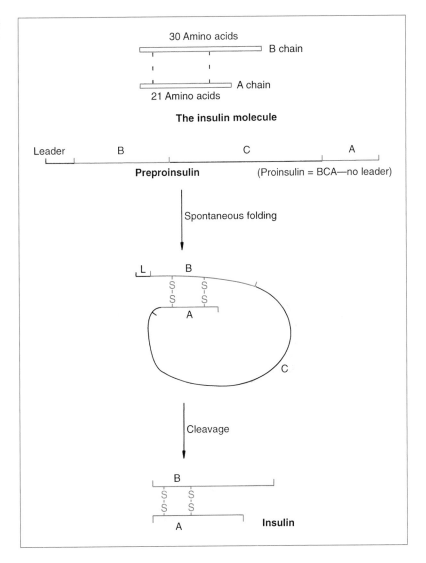

expressed as fusion proteins, consisting of the first few amino acids of β-galactosidase followed by the A or B polypeptides (Figure 14.2(b)). Each gene was designed so that its β-galactosidase and insulin segments were separated by a methionine residue, so that the insulin polypeptides could be cleaved from the β-galactosidase segments by treatment with cyanogen bromide (p. 281). The purified A and B chains were then attached to each other by disulphide bond formation in the test tube.

The final step, involving disulphide bond formation, is actually rather inefficient. A subsequent improvement has been to synthesize not the individual A and B genes, but the entire proinsulin reading frame, specifying B chain-C chain-A chain (Figure 14.1). Although this is a more daunting proposition in terms of DNA synthesis, the prohormone has the big advantage of folding

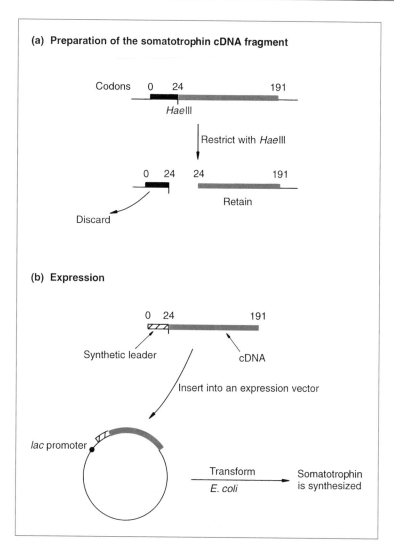

(a) Preparation of the somatotrophin cDNA fragment

(b) Expression

Figure 14.4 Production of recombinant somatotrophin.

Recombinant factor VIII

Although a number of important pharmaceutical compounds have been obtained from genes cloned in *E. coli*, the general problems associated with using bacteria to synthesize foreign proteins (p. 282) have led in many cases to these organisms being replaced by eukaryotes. An example of a recombinant pharmaceutical produced in eukaryotic cells is human factor VIII, a protein that plays a central role in blood clotting. The commonest form of haemophilia in humans results from an inability to synthesize factor VIII, leading to a breakdown in the blood clotting pathway and the well known symptoms associated with the disease.

Until recently the only way to treat haemophilia was by injection of purified factor VIII protein, obtained from human blood provided by donors. Purification of factor VIII is a complex procedure and the treatment is very

Figure 14.5 The factor VIII gene and its translation product.

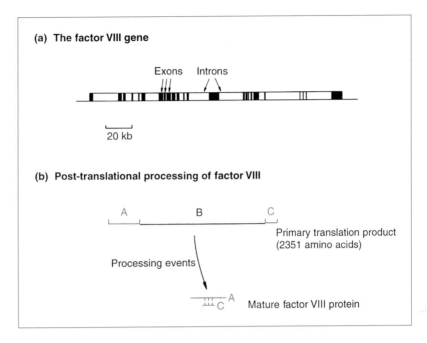

(a) **The factor VIII gene**

Exons Introns

20 kb

(b) **Post-translational processing of factor VIII**

A   B   C

Primary translation product
(2351 amino acids)

Processing events

A
C

Mature factor VIII protein

expensive. More critically, the purification is beset with difficulties, in particular in removing virus particles that may be present in the blood. Hepatitis and acquired immune deficiency syndrome (AIDS) can be and have been passed on to haemophiliacs via factor VIII injections. Recombinant factor VIII, free from contamination problems, would be a significant achievement for biotechnology.

The factor VIII gene is very large, over 186 kb in length, and is split into 26 exons and 25 introns (Figure 14.5(a)). The mRNA codes for a large polypeptide (2351 amino acids) which undergoes a complex series of post-translational processing events, eventually resulting in a dimeric protein consisting of a large subunit, derived from the upstream region of the initial polypeptide, and a small subunit from the downstream segment (Figure 14.5(b)). The two subunits contain a total of 17 disulphide bonds and a number of glycosylated sites. As might be anticipated for such a large and complex protein, it has not been possible to synthesize an active version in *E. coli*.

Initial attempts to obtain recombinant factor VIII therefore involved mammalian cells. In the first experiments to be carried out the entire cDNA was cloned in hamster cells, but yields of protein were disappointingly low. This was probably because the post-translational events, although carried out correctly in hamster cells, do not convert all of the initial product into an active form, limiting the overall yield. As an alternative, two separate fragments from the cDNA were used, one fragment coding for the large subunit polypeptide, the second for the small subunit. Each cDNA fragment was

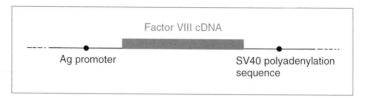

Figure 14.6 The expression signals used in production of recombinant factor VIII. The promoter is an artificial hybrid of the chicken β-actin and rabbit β-globin sequences, and the polyadenylation signal (needed for correct processing of the mRNA before translation into protein) is obtained from SV40 virus.

ligated into an expression vector, downstream of the Ag promoter (a hybrid between the chicken β-actin and rabbit β-globin sequences) and upstream of a polyadenylation signal from SV40 virus (Figure 14.6). The plasmid was introduced into a hamster cell line and recombinant protein obtained. The yields were over ten times greater than those from cells containing the complete cDNA, and the resulting factor VIII protein was indistinguishable in terms of function from the native form.

The most recent technology for production of recombinant factor VIII involves pharming (p. 289). The complete human cDNA has been attached to the promoter for the whey acidic protein gene of pig, leading to synthesis of human factor VIII in pig mammary tissue and subsequent secretion of the protein in the milk. The factor VIII produced in this way appears to be exactly the same as the native protein and is fully functional in blood clotting assays.

14.1.4 Synthesis of other recombinant human proteins

The list of human proteins synthesized by recombinant technology continues to grow (Table 14.1). As well as proteins used to treat disorders by replacement or supplementation of the malfunctional versions, the list includes a number of growth factors (e.g. interferons and interleukins) with potential uses in cancer therapy. These proteins are synthesized in very limited amounts in the body, so recombinant technology is the only viable means of obtaining them in the quantities needed for clinical purposes. Other proteins, such as serum albumin, are more easily obtained, but are needed in such large quantities that production in microorganisms is still a more attractive option.

14.1.5 Recombinant vaccines

The final category of recombinant protein is slightly different from the examples given in Table 14.1. A vaccine is an antigenic preparation that, after injection into the bloodstream, stimulates the immune system to synthesize

Table 14.1 Some of the human proteins that have been synthesized from genes cloned in bacteria and/or eukaryotic cells or by pharming.

Protein	Used in the treatment of
Insulin	Diabetes
Somatostatin	Growth disorders
Somatotrophin	Growth disorders
Factor VIII	Haemophilia
Factor IX	Christmas disease
Interferon-α	Leukaemia and other cancers
Interferon-β	Cancers, AIDS
Interferon-γ	Cancers, rheumatoid arthritis
Interleukins	Cancers, immune disorders
Granulocyte colony stimulating factor	Cancers
Tumour necrosis factor	Cancers
Epidermal growth factor	Ulcers
Fibroblast growth factor	Ulcers
Erythropoietin	Anaemia
Tissue plasminogen activator	Heart attack
Superoxide dismutase	Free radical damage in kidney transplants
Lung surfactant protein	Respiratory distress
α_1-antitrypsin	Emphysema
Serum albumin	Used as a plasma supplement
Relaxin	Used to aid childbirth
Deoxyribonuclease	Cystic fibrosis

antibodies that protect the body against infection. The antigenic material present in a vaccine is normally an inactivated form of the infectious agent. For example, antiviral vaccines often consist of virus particles that have been attenuated by heating or a similar treatment. In the past, two problems have hindered the preparation of attenuated viral vaccines.

(1) The inactivation process must be 100% efficient, as the presence in a vaccine of just one live virus particle could result in infection. This has been a problem with vaccines for the cattle disease foot-and-mouth.

(2) The large amounts of virus particles needed for vaccine production are usually obtained from tissue cultures. Unfortunately some viruses, notably hepatitis B virus, do not grow in tissue culture.

Producing vaccines as recombinant proteins

The use of gene cloning in this field centres on the discovery that virus-specific antibodies are sometimes synthesized in response not only to the whole virus particle, but also to isolated components of the virus. This is particularly true of purified preparations of the proteins present in the virus coat (Figure 14.7). If the genes coding for the antigenic proteins of a particular

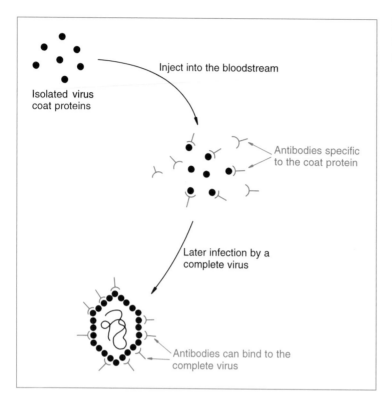

Figure 14.7 The principle behind the use of a preparation of isolated virus coat proteins as a vaccine.

Isolated virus coat proteins

Inject into the bloodstream

Antibodies specific to the coat protein

Later infection by a complete virus

Antibodies can bind to the complete virus

virus could be identified and inserted into an expression vector, the methods described above for the synthesis of animal proteins could be employed in the production of recombinant proteins that might be used as vaccines. These vaccines would have the advantages that they would be free of intact virus particles and they could be obtained in large quantities.

Unfortunately this approach has not been entirely successful, mainly because it has turned out that recombinant coat proteins often lack the full antigenic properties of the intact virus. The one notable success has been with hepatitis B virus, whose coat protein (the 'major surface antigen') has been synthesized in *Saccharomyces cerevisiae*, using a vector based on the 2 μm plasmid (p. 131). The protein was obtained in reasonably high quantities, and when injected into monkeys it provided protection against hepatitis B. This recombinant vaccine has been approved for use in humans.

Live recombinant vaccines

The use of live vaccinia virus as a vaccine for smallpox dates back to 1796, when Edward Jenner first realized that this virus, harmless to humans, could stimulate immunity against the much more dangerous smallpox virus. The term 'vaccine' comes from vaccinia; its use resulted in the worldwide eradication of smallpox in 1980.

Figure 14.8 The rationale behind the potential use of a recombinant vaccinia virus.

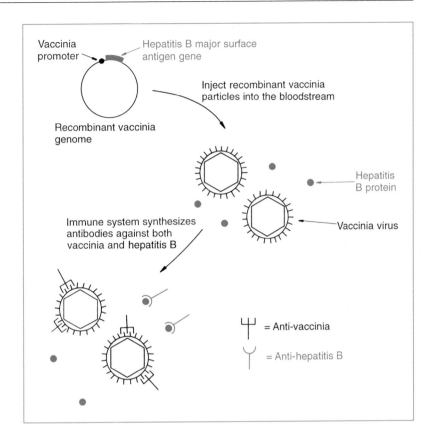

A more recent idea is that recombinant vaccinia viruses could be used as live vaccines against other diseases. If a gene coding for a virus coat protein, for example the hepatitis B major surface antigen, is ligated into the vaccinia genome under control of a vaccinia promoter, then the gene will be expressed (Figure 14.8). After injection into the bloodstream, replication of the recombinant virus results not only in new vaccinia particles, but also in significant quantities of the major surface antigen. Immunity against both smallpox and hepatitis B would result.

This remarkable technique has considerable potential. Recombinant vaccinia viruses expressing a number of foreign genes have been constructed and shown to confer immunity against the relevant diseases in experimental animals (Table 14.2). The possibility of broad-spectrum vaccines is raised by the demonstration that a single recombinant vaccinia, expressing the genes for influenza virus haemagglutinin, hepatitis B major surface antigen and herpes simplex virus glycoprotein, confers immunity against each disease in monkeys. The important question that must now be answered is whether we know enough about viral biology to be sure that release of recombinant vaccinia viruses into the biosphere via vaccination programmes can be allowed.

Table 14.2 Some of the foreign genes that have been expressed in recombinant vaccinia viruses.

Gene
Plasmodium falciparum (malaria parasite) surface antigen
Influenza virus coat proteins
Rabies virus G protein
Hepatitis B major surface antigen
Herpes simplex glycoproteins
Human immunodeficiency virus (HIV) envelope proteins
Vesicular somatitis coat proteins
Sindbis virus proteins

Table 14.3 Some of the commonest genetic diseases in the UK.

Disease	Symptoms	Frequency (births per year)
Inherited breast cancer	Cancer	1 in 300 females
Cystic fibrosis	Lung disease	1 in 2000
Huntington's chorea	Neurodegeneration	1 in 2000
Duchenne muscular dystrophy	Progressive muscle weakness	1 in 3000 males
Haemophilia A	Blood disorder	1 in 4000 males
Sickle cell anaemia	Blood disorder	1 in 10000
Phenylketonuria	Mental retardation	1 in 12000
β-thalassaemia	Blood disorder	1 in 20000
Retinoblastoma	Cancer of the eye	1 in 20000
Haemophilia B	Blood disorder	1 in 25000 males
Tay-Sachs disease	Blindness, loss of motor control	1 in 200000

14.2 Identification of genes responsible for human diseases

A second major area of medical research in which gene cloning is having an impact is in the identification and isolation of genes responsible for human diseases. A genetic or inherited disease is one that is caused by a defect in a specific gene (Table 14.3), individuals carrying the defective gene being predisposed towards developing the disease at some stage of their lives. With some inherited diseases, such as haemophilia, the gene is present on the X chromosome, so all males carrying the gene express the disease state; females with one defective gene and one correct gene are healthy but can pass the disease on to their male offspring. Genes for other diseases are present on autosomes and in most cases are recessive, so both chromosomes of the pair must carry a defective version for the disease to occur; a few diseases, including Huntington's chorea, are autosomal dominant, so a single copy of the defective gene is enough to lead to the disease state.

With some genetic diseases, the symptoms manifest themselves early in life, with others the disease may not be expressed until the individual is middle-aged or elderly. Cystic fibrosis is an example of the former, neuro-degenerative diseases such as Alzheimer's and Huntington's are examples of the latter. With a number of diseases that appear to have a genetic component, cancers in particular, the overall syndrome is complex with the disease remaining dormant until triggered by some metabolic or environmental stimulus. If predisposition to these diseases can be diagnosed, the risk factor can be reduced by careful management of the patient's lifestyle to minimize the chances of the disease being triggered.

Genetic diseases have always been present in the human population but their importance has increased in recent decades. This is because vaccination programmes, antibiotics and improved sanitation have reduced the prevalence of infectious diseases such as smallpox, tuberculosis and cholera, which were major killers of the early twentieth century. The result is that a greater proportion of the population now dies from a disease that has a genetic component, especially the late-onset diseases that are now more common because of increased life expectancies. Medical research has been successful in controlling many infectious diseases; can it be equally successful with genetic disease?

There are a number of reasons why identifying the gene responsible for a genetic disease is important.

(1) Gene identification may provide an indication of the biochemical basis to the disease, enabling therapies to be designed.

(2) Identification of the mutation present in a defective gene can be used to devise a screening programme so that the mutant gene can be identified in individuals who are carriers or who have not yet developed the disease. Carriers can receive counselling regarding the chances of their children inheriting the disease. Early identification in individuals who have not yet developed the disease allows appropriate precautions to be taken to reduce the risk of the disease becoming expressed.

(3) Identification of the gene is a prerequisite for gene therapy (p. 309).

How to identify a gene for a genetic disease

14.2.1

There is no single strategy for identification of genes that cause diseases, the best approach depending on the information that is available about the disease. To gain an understanding of the principles of this type of work we will consider the most common and most difficult scenario. This is when all that is known about the disease is that certain people have it. Even with such an unpromising starting point, DNA techniques can locate the relevant gene.

Locating the approximate position of the gene in the human genome

If there is no information about the desired gene, how can it be located in the human genome sequence? The answer is to return to basic genetic principles in order to determine the approximate position of the gene on the human genetic map. Genetic mapping is usually carried out by **linkage analysis**, which involves comparing the inheritance pattern for the target gene with the inheritance patterns for genetic loci whose map positions are already known. If two loci are inherited together they must be very close on the same chromosome: if this is not the case then recombination events and the random segregation of chromosomes during meiosis will result in the loci displaying different inheritance patterns (Figure 14.9). Demonstration of linkage with one or more mapped genetic loci is therefore the key to understanding the chromosomal position of an unmapped gene.

In humans it is not possible to carry out directed breeding programmes aimed at determining the map position of a desired gene. Instead, mapping of disease genes must make use of data available from **pedigree analysis**, in which inheritance of the gene is examined in families with a high incidence of the disease being studied. It is important to be able to obtain DNA samples from at least three generations of each family, and the more family members there are the better, but unless the disease is very uncommon it is usually possible to find suitable pedigrees. Linkage between the presence/absence of the disease and the inheritance of other genes could be studied, but as DNA samples are being analysed, linkage to DNA markers is more usually tested (p. 255).

To illustrate how linkage analysis is used we will look briefly at the way in which one of the genes for human breast cancer was mapped. The first breakthrough in this project occurred in 1990 as a result of **restriction fragment length polymorphism (RFLP) linkage analyses** carried out by a group at the University of California at Berkeley. This study showed that in families with a high incidence of breast cancer, a significant number of the women who suffered from the disease all possessed the same version of an RFLP called *D17S74*. This RFLP had previously been mapped to the long arm of chromosome 17 (Figure 14.10): the gene being sought – *BRCA1* – must therefore also be located on the long arm of chromosome 17. This initial linkage result was extremely important, as it indicated in which region of the human genome the breast cancer gene was to be found, but it was far from the end of the story. In fact, over 1000 genes are thought to lie in this particular 20 Mb stretch of chromosome 17. The next objective was therefore to carry out more linkage studies to try to pinpoint *BRCA1* more accurately.

Figure 14.9 Inheritance patterns for linked and unlinked genes. Three families are shown, circles representing females and squares representing males. (a) Two closely linked genes are almost always inherited together. (b) Two genes on different chromosomes display random segregation. (c) Two genes that are far apart on a single chromosome are often inherited together, but recombination may unlink them.

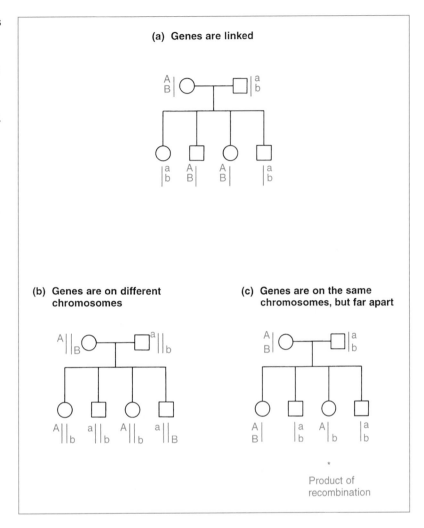

This was achieved by first examining the region containing *BRCA1* for short tandem repeats (STRs) (p. 255), which are useful for fine scale mapping because many of them exist in three or more allelic forms, rather than just the two alleles that are possible for an RFLP. Several alleles of an STR might therefore be present within a single pedigree, enabling more detailed mapping to be carried out. Short tandem repeat linkage mapping reduced the size of the *BRCA1* region from 20 Mb down to jut 600 kb (Figure 14.10). This approach to locating a gene is called **positional cloning**.

Identification of candidates for the disease gene

One might imagine that once the map location of the disease gene has been determined the next step is simply to refer to the genome sequence in order to identify the gene. Unfortunately a great deal of work still has to be done. Genetic mapping, even in its most precise form, only gives an approximate

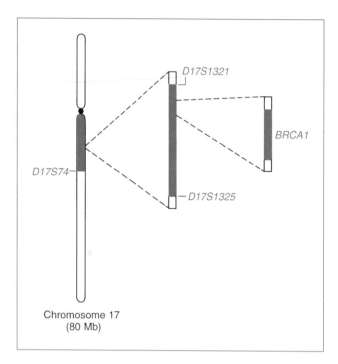

Figure 14.10 Mapping the breast cancer gene. Initially the gene was mapped to a 20 Mb segment of chromosome 17 (highlighted region in the left drawing). Additional mapping experiments narrowed this down to a 600 kb region flanked by two previously mapped loci, *D17S1321* and *D17S1325* (middle drawing). After examination of expressed sequences, a strong candidate for *BRCA1* was eventually identified (right drawing).

indication of the location of the gene. In the breast cancer project the researchers were fortunate in being able to narrow the search area down to just 600 kb; often 10 Mb or more of DNA sequence has to be examined. Such large stretches of DNA could contain many genes: the 600 kb breast cancer region contained over 60 genes, any one of which could have been *BRCA1*.

A variety of approaches can be used to identify which of the genes in the mapped region is the disease gene.

(1) The expression profiles of the **candidate genes** can be examined by hybridization analysis or reverse transcription–polymerase chain reaction (RT–PCR) (p. 224) of RNA from different tissues. For example, *BRCA1* would be expected to hybridize to RNA prepared from breast tissue, and also to ovary tissue RNA, ovarian cancer frequently being associated with inherited breast cancer.

(2) Southern hybridization analysis (p. 198) can be carried out with DNA from different species (these are called **zoo blots**). The rationale is that an important human gene will almost certainly have homologues in other mammals, and that this homologue, although having a slightly different sequence from the human version, will be detectable by hybridization with a suitable probe.

(3) The gene sequences could be examined in individuals with and without the disease to see if the genes from affected individuals contain mutations that might explain why they have the disease.

(4) To confirm the identity of a candidate gene, it might be possible to prepare a knockout mouse (p. 263) that has an inactive version of the equivalent mouse gene. If the knockout mouse displays symptoms compatible with the human disease then the candidate gene is almost certainly the correct one.

When applied to the breast cancer region, these analyses resulted in identification of an approximately 100kb gene, made up of 22 exons and coding for a 1863 amino acid protein, that is a strong candidate for *BRCA1*. Transcripts of the gene were detectable in breast and ovary tissues, and homologues are present in mice, rats, rabbits, sheep and pigs, but not chickens. Most importantly, the genes from five susceptible families contained mutations (such as frameshift and nonsense mutations) likely to lead to a non-functioning protein. Although circumstantial, the evidence in support of the candidate was sufficiently overwhelming for this gene to be identified as *BRCA1*.

14.3 Gene therapy

The final application of recombinant DNA technology in medicine that we will consider is gene therapy. This is the name originally given to methods that aim to cure an inherited disease by providing the patient with a correct copy of the defective gene. Gene therapy has now been extended to include attempts to cure any disease by introduction of a cloned gene into the patient. First we will examine the techniques used in gene therapy, and then we will attempt to address the ethical issues.

14.3.1 Gene therapy for inherited diseases

There are two basic approaches to gene therapy: germline therapy and somatic cell therapy. In germline therapy a fertilized egg is provided with a copy of the correct version of the relevant gene and reimplanted into the mother. If successful, the gene is present and expressed in all cells of the resulting individual. Germline therapy is usually carried out by microinjection of DNA into the isolated egg cell (p. 103), and theoretically could be used to treat any inherited disease.

Somatic cell therapy involves manipulation of ordinary cells, usually ones which can be removed from the organism, transfected, and then placed back in the body. The technique has most promise for inherited blood diseases (e.g. haemophilia and thalassaemia), with genes being introduced into stem cells from the bone marrow, which give rise to all the specialized cell types in the blood. The strategy is to prepare a bone extract containing several billion cells, transfect these with a retrovirus-based vector, and then reimplant the cells. Subsequent replication and differentiation of transfectants

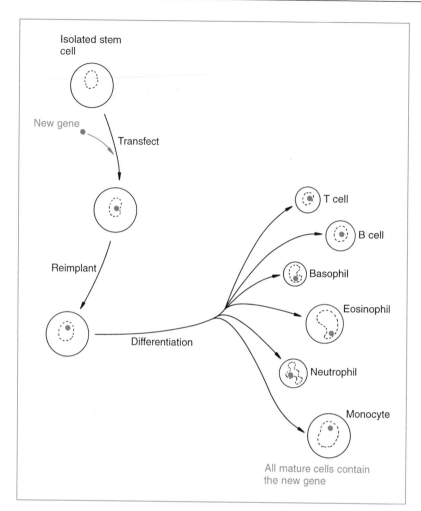

Figure 14.11 Differentiation of a transfected stem cell leads to the new gene being present in all the mature blood cells.

Isolated stem cell

New gene

Transfect

Reimplant

Differentiation

T cell

B cell

Basophil

Eosinophil

Neutrophil

Monocyte

All mature cells contain the new gene

leads to the added gene being present in all the mature blood cells (Figure 14.11). The advantage of a retrovirus vector is that this type of vehicle has an extremely high transfection frequency, enabling a large proportion of the stem cells in a bone marrow extract to receive the new gene.

Somatic cell therapy also has potential in the treatment of lung diseases such as cystic fibrosis, as DNA cloned in adenovirus vectors (p. 153) or contained in liposomes (p. 147) is taken up by the epithelial cells in the lungs after introduction into the respiratory tract via an inhaler. However, gene expression occurs for only a few weeks, and as yet this has not been developed into an effective means of treating cystic fibrosis.

With those genetic diseases where the defect arises because the mutated gene does not code for a functional protein, all that is necessary is to provide the cell with the correct version of the gene: removal of the defective genes is unnecessary. The situation is less easy with dominant genetic diseases

(p. 305), as with these it is the defective gene product itself that is responsible for the disease state, and so the therapy must include not only addition of the correct gene but also removal of the defective version. This requires a gene delivery system that promotes recombination between the chromosomal and vector-borne versions of the gene, so that the defective chromosomal copy is replaced by the gene from the vector. The technique is complex and unreliable, and broadly applicable procedures have not yet been developed.

14.3.2   Gene therapy and cancer

The clinical uses of gene therapy are not limited to treatment of inherited diseases. There have also been attempts to use gene cloning to disrupt the infection cycles of human pathogens such as the AIDS virus. However, the most intensive area of current research into gene therapy concerns its potential use as a treatment for cancer.

Most cancers result from activation of an oncogene that leads to tumour formation, or inactivation of a gene that normally suppresses formation of a tumour. In both cases a gene therapy could be envisaged to treat the cancer. Introduction of a gene for an **antisense RNA** copy (p. 318) of an oncogene could, for example, reduce or prevent expression of the oncogene and reverse its tumorigenic activity. If the cancer is caused by inactivation of a tumour suppressor gene, the gene therapy would involve introduction of an active version of that gene. The stumbling block at the moment is not in identification of appropriate genes to use in cancer gene therapy, but in devising suitable delivery methods that ensure that the cloned gene is taken up by the cancerous cells.

With cancers, a novel application of gene therapy would be to introduce a gene that selectively kills cancer cells. This is looked on as the most effective and general approach to the treatment of many types of cancer, mainly because it does not require a detailed understanding of the genetic basis of the particular disease being treated. Many genes that code for toxic proteins are known, and introduction of one of these into a tumour should result in the death of the cancer cells and recovery of the patient. Clearly the key factor is that the cloned gene must be targeted specifically at the cancer cells, so that healthy cells are not killed. This would require a very accurate delivery system or some means of ensuring that the gene is expressed only in the cancer cells, for example by placing it under control of a promoter that is active only in those cells.

Another approach is to use gene therapy to improve the natural killing of cancer cells by the patient's immune system, perhaps with a gene that causes the tumour cells to synthesize strong antigens that are efficiently recognized by the immune system. All of these approaches, and many not based on gene therapy, are currently being tested in the fight against cancer.

14.3.3

The ethical issues raised by gene therapy

Should gene therapy be used to cure human disease? As with many ethical questions there is no simple answer. On the one hand, there could surely be no justifiable objection to the routine application via a respiratory inhaler of correct versions of the cystic fibrosis gene as a means of managing this disease. Similarly, if bone marrow transplants are acceptable, then it is difficult to argue against gene therapies aimed at correction of blood disorders via stem cell transfection. And cancer is such a terrible disease that the withholding of effective treatment regimens on moral grounds could itself be criticized as immoral.

Germline therapy is a more difficult issue. The problem is that the techniques used for germline correction of inherited diseases are exactly the same techniques that could be used for germline manipulation of other inherited characteristics. Indeed, the development of this technique with animals has not been prompted by any desire to cure genetic diseases, the aims being to 'improve' farm animals, for example by making genetic changes that result in lower fat content. This type of manipulation, where the genetic constitution of an organism is changed in a directed, heritable fashion, is clearly unacceptable with humans. At present, technical problems mean that human germline manipulation would be difficult. Before these problems are solved we should ensure that the desire to do good should not raise the possibility of doing tremendous harm.

Further reading

Boucher, R.C. (1999) Status of gene therapy for cystic fibrosis lung disease. *Journal of Clinical Investigation*, **103**, 441–5.

Broder, C.C. & Earl, P.L. (1999) Recombinant vaccinia viruses – design, generation, and isolation. *Molecular Biotechnology*, **13**, 223–45.

Goeddel, D.V., Heyneker, H.L., Hozumi, T. *et al.* (1979) Direct expression in *Escherichia coli* of a DNA sequence coding for human growth hormone. *Nature*, **281**, 544–8. [Production of recombinant somatotrophin.]

Goeddel, D.V., Kleid, D.G., Bolivar, F. *et al.* (1979) Expression in *Escherichia coli* of chemically synthesized genes for human insulin. *Proceedings of the National Academy of Sciences of the USA*, **76**, 106–10.

Itakura, K., Hirose, T., Crea, R. *et al.* (1977) Expression in *Escherichia coli* of a chemically synthesized gene for the hormone somatostatin. *Science*, **198**, 1056–63.

Lemoine, N. & Cooper, D. (1998) *Gene Therapy*. BIOS Scientific Publishers, Oxford.

Liu, M.A. (1998) Vaccine developments. *Nature Medicine*, **4**, 515–19. [Describes the current position with development of recombinant vaccines.]

Miki, Y., Swensen, J., Shattuck Eidens, D. *et al.* (1994) A strong candidate for the breast and ovarian cancer susceptibility gene *BRCA1*. *Science*, **266**, 66–71.

Paleyanda, R.K., Velander, W.H., Lee, T.K. *et al.* (1997) Transgenic pigs produce functional human factor VIII in milk. *Nature Biotechnology*, **15**, 971–5.

Strachan, T. & Read, A.P. (1999) *Human Molecular Genetics*, 2nd edn. BIOS Scientific Publishers, Oxford. [Contains particularly good descriptions of the identification of disease genes and gene therapy.]

Chapter 15   Gene Cloning and DNA Analysis in Agriculture

Agriculture, or more specifically the cultivation of plants, is the world's oldest biotechnology, with an unbroken history that stretches back at least 10 000 years. Throughout this period humans have constantly searched for improved varieties of their crop plants: varieties with better nutritional qualities, higher yields, or features that aid cultivation and harvesting. During the first few millennia, crop improvements occurred in a sporadic fashion, but in recent centuries new varieties have been obtained by breeding programmes of ever increasing sophistication. However, the most sophisticated breeding programme still retains an element of chance, dependent as it is on the random merging of parental characteristics in the hybrid offspring that are produced. The development of a new variety of crop plant, displaying a precise combination of desired characteristics, is a lengthy and difficult process.

Gene cloning provides a new dimension to crop breeding by enabling directed changes to be made to the genotype of a plant, circumventing the random processes inherent in conventional breeding. Two general strategies have been used:

(1) Gene addition, in which cloning is used to alter the characteristics of a plant by providing it with one or more new genes.
(2) Gene subtraction, in which genetic engineering techniques are used to inactivate one or more of the plant's existing genes.

A number of projects are being carried out around the world, many by biotechnology companies, aimed at exploiting the potential of gene addition and gene subtraction in crop improvement. In this chapter we will investigate a representative selection of these projects, and look at some of the problems that must be solved if plant genetic engineering is to gain widespread acceptance in agriculture.

15.1

The gene addition approach to plant genetic engineering

Gene addition involves the use of cloning techniques to introduce into a plant one or more new genes coding for a useful characteristic that the plant lacks. A good example of the technique is provided by the development of plants that resist insect attack by synthesizing insecticides coded by cloned genes.

15.1.1

Plants that make their own insecticides

Plants are subject to predation by virtually all other types of organism – viruses, bacteria, fungi and animals – but in agricultural settings the greatest problems are caused by insects. To reduce losses, crops are regularly sprayed with insecticides. Most conventional insecticides (e.g. pyrethroids and organophosphates) are relatively non-specific poisons that kill a broad spectrum of insects, not just the ones eating the crop. Because of their high toxicity, several of these insecticides also have potentially harmful side effects for other members of the local biosphere, including in some cases humans. These problems are exacerbated by the need to apply conventional insecticides to the surfaces of plants by spraying, which means that subsequent movement of the chemicals in the ecosystem cannot be controlled. Furthermore, insects that live within the plant, or on the undersurfaces of leaves, can sometimes avoid the toxic effects altogether.

What features would be displayed by the ideal insecticide? Clearly it must be toxic to the insects against which it is targeted, but if possible this toxicity should be highly selective, so that the insecticide is harmless to other insects and is not poisonous to animals and to humans. The insecticide should be biodegradable, so that any residues that remain after the crop is harvested, or which are carried out of the field by rainwater, do not persist long enough to damage the environment. And it should be possible to apply the insecticide in such a way that all parts of the crop, not just the upper surfaces of the plants, are protected against insect attack.

The ideal insecticide has not yet been discovered. The closest we have are the δ-endotoxins produced by the soil bacterium *Bacillus thuringiensis*.

The δ-endotoxins of *Bacillus thuringiensis*

Insects not only eat plants: bacteria also form an occasional part of their diet. In response, several types of bacteria have evolved defence mechanisms against insect predation, an example being *B. thuringiensis* which, during sporulation, forms intracellular crystalline bodies that contain an insecticidal protein called the δ-endotoxin. The activated protein is highly poisonous to insects, some 80000 times more toxic than organophosphate insecticides, and is relatively selective, different strains of the bacterium synthesizing proteins effective against the larvae of different groups of insects (Table 15.1).

Table 15.1 The range of insects poisoned by the various types of *B. thuringiensis* δ-endotoxins.

δ-endotoxin type	Effective against
CryI	Lepidoptera (moth and butterfly) larvae
CryII	Lepidoptera and Diptera (two-winged fly) larvae
CryIII	Lepidoptera larvae
CryIV	Diptera larvae
CryV	Nematode worms
CryVI	Nematode worms

The δ-endotoxin protein that accumulates in the bacterium is an inactive precursor. After ingestion by the insect this protoxin is cleaved by proteases, resulting in shorter versions of the protein that display the toxic activity, by binding to the inside of the insect's gut and damaging the surface epithelium, so that the insect is unable to feed and consequently starves to death (Figure 15.1). Variation in the structure of these binding sites in different groups of insects is probably the underlying cause of the high specificities displayed by the different types of δ-endotoxin.

B. thuringiensis toxins are not recent discoveries, the first patent for their use in crop protection having been granted in 1904. Over the years there have been several attempts to market them as environmentally friendly insecticides, but their biodegradability acts as a disadvantage because it means that they must be reapplied at regular intervals during the growing season, increasing the farmer's costs. Current research is therefore aimed at developing δ-endotoxins that do not require regular application. One approach is via protein engineering (p. 214), modifying the structure of the toxin so that it is more stable. A second approach is to engineer the crop to synthesize its own toxin.

Cloning a δ-endotoxin gene in maize

Maize is an example of a crop plant that is not served well by conventional insecticides. A major pest is the European corn borer (*Ostrinia nubilialis*), which tunnels into the plant from eggs laid on the undersurfaces of leaves, thereby evading the effects of insecticides applied by spraying. The first attempt at countering this pest by engineering maize plants to synthesize δ-endotoxin was made by plant biotechnologists at a Ciba-Geigy laboratory in North Carolina. They worked with the CryIA(b) version of the toxin, which had previously been shown to be a 1155 amino acid protein, with the toxic activity residing in the segment from amino acids 29–607. Rather than isolating the natural gene, the Ciba-Geigy group made a shortened version, containing the first 648 codons, by artificial gene synthesis. This strategy enabled them to introduce modifications into the gene to improve its

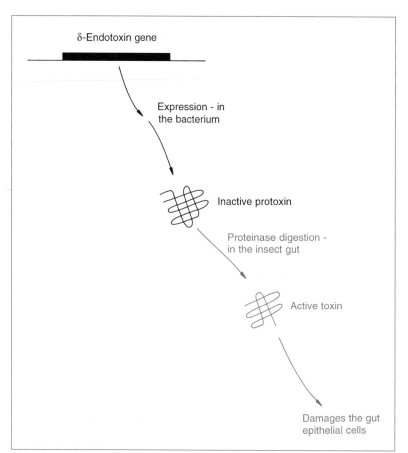

Figure 15.1 Mode of action of a δ-endotoxin.

expression in maize plants. For example, the codons that were used in the artificial gene were those known to be preferred by maize, and the overall GC content of the gene was set at 65%, compared with the 38% GC content of the native bacterial version of the gene (Figure 15.2(a)). The artificial gene was ligated into a cassette vector (p. 279) between a promoter and polyadenylation signal from cauliflower mosaic virus (Figure 15.2(b)), and introduced into maize embryos by bombardment with DNA-coated micro-projectiles (p. 103). The embryos were grown into mature plants, and trans-formants identified by polymerase chain reaction (PCR) analysis of DNA extracts, using primers specific for a segment of the artificial gene (Figure 15.2(c)).

The next step was to use an immunological test to determine if δ-endotoxin was being synthesized by the transformed plants. The results showed that the artificial gene was indeed active, but that the amounts of δ-endotoxin being produced varied from plant to plant, from about 250 to 1750 ng of toxin per mg of total protein. These differences were probably due to **positional effects**, the level of expression of a gene cloned in a plant (or

Figure 15.2 Important steps in the procedure used to obtain genetically engineered maize plants expressing an artificial δ-endotoxin gene.

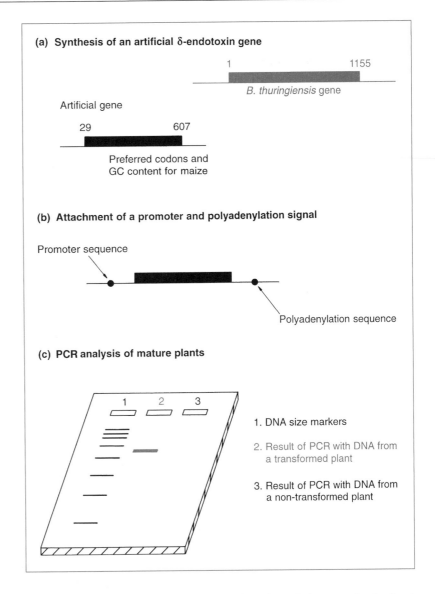

(a) Synthesis of an artificial δ-endotoxin gene

1   1155

B. thuringiensis gene

Artificial gene

29   607

Preferred codons and
GC content for maize

(b) Attachment of a promoter and polyadenylation signal

Promoter sequence

Polyadenylation sequence

(c) PCR analysis of mature plants

1 2 3

1. DNA size markers

2. Result of PCR with DNA from a transformed plant

3. Result of PCR with DNA from a non-transformed plant

animal) often being influenced by the exact location of the gene in the host chromosomes (Figure 15.3).

Were the transformed plants able to resist the attentions of the corn borers? This was assessed by field trials in which transformed and normal maize plants were artificially infested with larvae and the effects of predation measured over a period of 6 weeks. Two criteria were used: (1) the amount of damage suffered by the foliage of the infested plants, and (2) the lengths of the tunnels produced by the larvae boring into the plants. In both respects the transformed plants gave better results than the normal ones. In particular, the average length of the larval tunnels was reduced from 40.7 cm for the controls to just 6.3 cm for the engineered plants. In real terms, this is a very significant level of resistance.

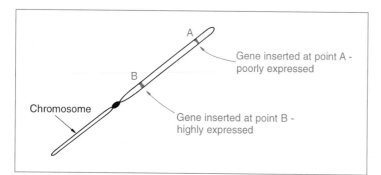

Figure 15.3 Positional effects.

Gene inserted at point A - poorly expressed

Gene inserted at point B - highly expressed

Chromosome

15.1.2 Other gene addition projects

Maize is not the only plant that has been engineered to produce δ-endotoxin, similar projects have been carried out with rice, cotton, potato, tomato and other crops. Nor is this the only approach to insect resistance. Equally successful results have been obtained with genes coding for proteinase inhibitors, small polypeptides that disrupt the activities of enzymes in the insect gut, preventing or slowing growth. Proteinase inhibitors are produced naturally by several types of plant, notably legumes such as cowpeas and common beans, and their genes have been successfully transferred to other crops which do not normally make significant amounts of these proteins. The inhibitors are particularly effective against beetle larvae that feed on seeds, and so may be a better alternative than δ-endotoxin for plants whose seeds are stored for long periods.

Examples of other gene addition projects are listed in Table 15.2. In many cases the objective is to improve the plant's ability to withstand pests such as insects, fungi, bacteria and viruses, or to be able to resist the toxic effects of herbicides used to control weeds. Other projects are starting to explore the use of genetic modification to improve the nutritional quality of crop plants, for example by increasing the content of essential amino acids and by changing the plant biochemistry so that more of the available nutrients can be utilized during digestion by humans or animals.

15.2 Gene subtraction

Gene subtraction is a misnomer, as the modification does not involve the actual removal of a gene, merely its inactivation. There are several ways by which a single, chosen gene could be inactivated in a living plant, the most successful so far in practical terms being the use of **antisense technology**. The example we will focus on is an important one, as it resulted in one of the first genetically modified foodstuffs to be approved for sale to the general public.

Table 15.2 Examples of gene addition projects with plants.

Gene for	Source organism	Characteristic conferred on modified plants
δ-endotoxin	*B. thuringiensis*	Insect resistance
Proteinase inhibitors	Various legumes	Insect resistance
Chitinase	Rice	Fungal resistance
Glucanase	Alfalfa	Fungal resistance
Ribosome-inactivating portein	Barley	Fungal resistance
Ornithine carbomyltransferase	*Pseudomonas syringae*	Bacterial resistance
RNA polymerase, helicase	Potato leafroll luteovirus	Virus resistance
Satellite RNAs	Various viruses	Virus resistance
Virus coat proteins	Various viruses	Virus resistance
2′–5′ oligoadenylate synthetase	Rat	Virus resistance
Acetolactate synthase	*Nicotiana tabacum*	Herbicide resistance
Enolpyruvylshikimate-3-phosphate synthase	*Agrobacterium* spp.	Herbicide resistance
Glyphosate oxidoreductase	*Ochrobactrum anthropi*	Herbicide resistance
Nitrilase	*Klebsiella ozaenae*	Herbicide resistance
Phosphinothricin acetyltransferase	*Streptomyces* spp.	Herbicide resistance
Barnase ribonuclease inhibitor	*Bacillus amyloliquefaciens*	Male sterility
DNA adenine methylase	*E. coli*	Male sterility
Methionine-rich protein	Brazil nuts	Improved sulphur content
Aminocyclopropane carboxylic acid deaminase	Various	Modified fruit ripening
S-Adenosylmethionine hydrolase	Bacteriophage T3	Modified fruit ripening
Monellin	*Thaumatococcus danielli*	Sweetness
Thaumatin	*T. danielli*	Sweetness
Acyl carrier protein thioesterase	*Umbellularia californica*	Modified fat/oil content
Delta-12 desaturase	*Glycine max*	Modified fat/oil content

15.2.1 The principle behind antisense technology

In an antisense experiment the gene to be cloned is ligated into the vector in reverse orientation (Figure 15.4). This means that when the cloned 'gene' is transcribed, the RNA that is synthesized is the reverse complement of the messenger RNA (mRNA) produced from the normal version of the gene. We refer to this reverse complement as an antisense RNA, sometimes abbreviated to asRNA.

An antisense RNA can prevent synthesis of the product of the gene it is directed against. The underlying mechanism is not altogether clear, but it almost certainly involves hybridization between the antisense and sense copies of the RNA (Figure 15.5). It is possible that the block to expression arises because the resulting double-stranded RNA molecule is rapidly degraded by cellular ribonucleases, or the explanation might be that the antisense RNA simply prevents ribosomes from attaching to the sense strand.

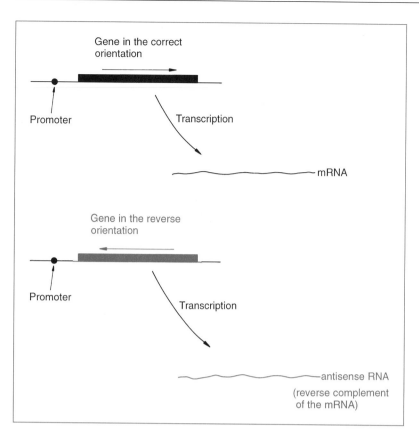

Figure 15.4 Antisense RNA.

Whatever the mechanism, synthesis of antisense RNA in a transformed plant is an effective way of carrying out gene subtraction.

15.2.2 Antisense RNA and the engineering of fruit ripening in tomato

At present, commercially grown tomatoes and other soft fruits are usually picked before they are completely ripe, to allow time for the fruits to be transported to the marketplace before they begin to spoil. This is essential if the process is to be economically viable, but there is a problem in that most immature fruits do not develop their full flavour if they are removed from the plant before they are fully ripe. The result is that mass-produced tomatoes often have a bland taste which makes them less attractive to the consumer. Two biotechnology companies – Calgene in the USA and ICI Seeds in the UK – used antisense technology as a means of genetically engineering tomato plants so that the fruit ripening process is slowed down. This enables the grower to leave the fruits on the plant until they ripen to the stage where the flavour has fully developed, there still being time to transport and market the crop before spoilage sets in.

Figure 15.5 Possible mechanisms for the inhibition of gene expression by antisense RNA.

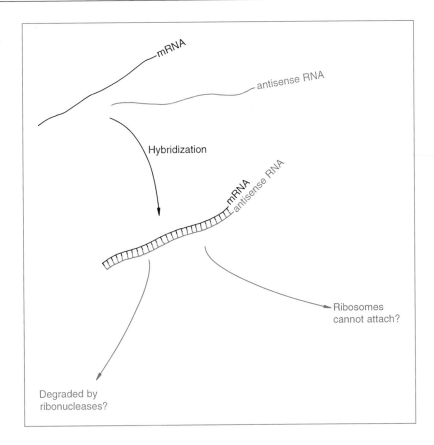

mRNA

antisense RNA

Hybridization

mRNA

antisense RNA

Ribosomes cannot attach?

Degraded by ribonucleases?

The role of the polygalacturonase gene in tomato fruit ripening

The timescale for development of a fruit is measured as the number of days or weeks after flowering. In tomato, this process takes approximately 8 weeks from start to finish, with the colour and flavour changes associated with ripening beginning after about 6 weeks. At about this time a number of genes involved in the later stages of ripening are switched on, including one coding for the polygalacturonase enzyme (Figure 15.6). This enzyme slowly breaks down the polygalacturonic acid component of the cell walls in the fruit pericarp, resulting in a gradual softening. The softening makes the fruit palatable, but if taken too far results in a squashy, spoilt tomato attractive only to students with limited financial resources.

Partial inactivation of the polygalacturonase gene might increase the time between flavour development and spoilage of the fruit. How could antisense technology be used to achieve this result?

Cloning the antisense polygalacturonase 'gene'

The experiment that we will follow was carried out by the Plant Biotechnology Section of ICI Seeds, together with scientists at the University of

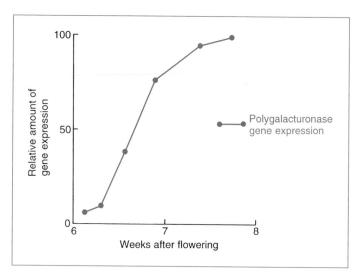

Figure 15.6 The increase in polygalacturonase gene expression seen during the later stages of tomato fruit ripening.

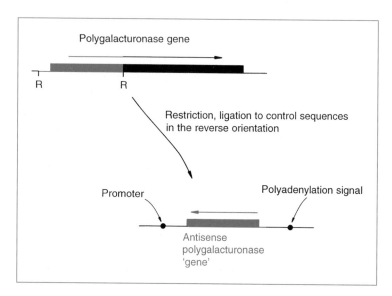

Figure 15.7 Construction of an antisense polygalacturonase 'gene'. R = restriction site.

Nottingham, in the mid-1980s. A 730 bp restriction fragment was obtained from the 5′ region of the normal polygalacturonase gene, representing just under half of the coding sequence (Figure 15.7). The orientation of the fragment was reversed and a cauliflower mosaic virus promoter ligated to the start of the sequence, and a plant polyadenylation signal attached to the end. The construction was then inserted into the Ti plasmid vector pBIN19 (p. 143). Once inside the plant, transcription from the cauliflower mosaic virus promoter should result in synthesis of an antisense RNA complementary to the first half of the polygalacturonase mRNA. Previous experiments with antisense RNA had suggested that this would be sufficient to reduce or even prevent translation of the target mRNA.

Figure 15.8 Obtaining transformed plants by infection of stem segments with recombinant *A. tumefaciens*.

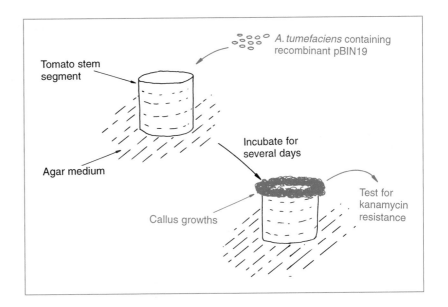

Transformation was carried out by introducing the recombinant pBIN19 molecules into *Agrobacterium tumefaciens* bacteria and then allowing the bacteria to infect tomato stem segments (Figure 15.8). Small amounts of callus material collected from the surfaces of these segments were tested for their ability to grow on an agar medium containing kanamycin (remember that pBIN19 carries a gene for kanamycin resistance; Figure 7.14). Resistant transformants were identified and allowed to develop into mature plants.

The results of the experiment were assessed in the following ways:

(1) The presence of the antisense 'gene' in the DNA of the transformed plants was checked by Southern hybridization (p. 198).

(2) Expression of the antisense 'gene' was measured by northern hybridization (p. 223) with a single-stranded DNA probe that would hybridize only to the antisense RNA.

(3) The effect of antisense RNA synthesis on the amount of polygalacturonase mRNA in the cells of ripening fruit was determined by northern hybridization with a second single-stranded DNA probe, this one specific for the sense mRNA. These experiments showed that ripening fruit from transformed plants contained less polygalacturonase mRNA than the fruits from normal plants.

(4) The amounts of polygalacturonase enzyme produced in the ripening fruits of transformed plants were estimated from the intensities of the relevant bands after separation of fruit proteins by polyacrylamide gel electrophoresis, and by directly measuring the enzyme activities in the fruits. The results showed that less enzyme was synthesized in transformed fruits (Figure 15.9).

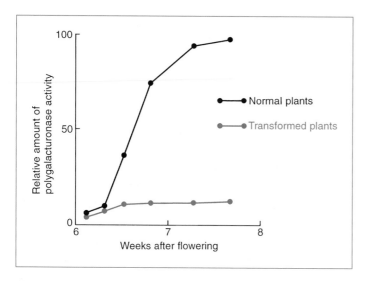

Figure 15.9 The differences in polygalacturonase activity in normal tomato fruits and in fruits expressing the antisense polygalacturonase 'gene'.

Table 15.3 Examples of gene subtraction projects with plants.

Target gene	Modified characteristic
Polygalacturonase	Delay of fruit spoilage in tomato
Aminocyclopropane carboxylic acid synthase	Modified fruit ripening in tomato
Polyphenol oxidase	Prevention of discoloration in fruits and vegetables
Starch synthase	Reduction of starch content in vegetables
Delta-12 desaturase	High oleic acid content in soybean
Chalcone synthase	Modification of flower colour in various decorative plants

Most importantly, the transformed fruits, although undergoing a gradual softening, could be stored for a prolonged period before beginning to spoil. This indicated that the antisense RNA had not completely inactivated the polygalacturonase gene, but had nonetheless produced a sufficient reduction in gene expression to delay the ripening process as desired.

15.2.3 Other examples of the use of antisense RNA in plant genetic engineering

In general terms, the applications of gene subtraction in plant genetic engineering are probably less broad than those of gene addition. It is easier to think of useful characteristics that a plant lacks and which might be introduced by gene addition, than it is to identify disadvantageous traits that the plant already possesses and which could be removed by gene subtraction. There are, however, a growing number of plant biotechnology projects based on gene subtraction (Table 15.3), and the approach is likely to increase in importance as the uncertainties that surround the underlying principles of antisense technology are gradually resolved.

15.3 Problems with genetically modified plants

Ripening-delayed tomatoes produced by gene subtraction were the first genetically modified whole food to be approved for marketing. Partly because of this, plant genetic engineering has provided the battleground on which biotechnologists and other interested parties have fought over the safety and ethical issues that arise from our ability to alter the genetic make-up of living organisms. A number of the most important questions do not directly concern genes and the expertise needed to answer them will not be found in this book. We cannot discuss in an authoritative fashion the possible impact, good or otherwise, that genetically modified crops might have on local farming practices in the developing world. Neither can we comment on the motives behind the development of herbicide-resistant plants by companies who also market the herbicide that farmers will have to use with the engineered crop. However we can, and should, look at the biological issues.

15.3.1 Safety concerns with selectable markers

One of the main areas of concern to emerge from the debate over genetically modified tomatoes is the possible harmful effects of the marker genes used with plant cloning vectors. Most plant vectors carry a copy of a gene for kanamycin resistance, enabling transformed plants to be identified during the cloning process. The kan^R gene, also called $nptII$, is bacterial in origin and codes for the enzyme neomycin phosphotransferase II. This gene and its enzyme product are present in all cells of an engineered plant. The fear that neomycin phosphotransferase might be toxic to humans has been allayed by tests with animal models, but two other safety issues remain:

(1) Could the kan^R gene contained in a genetically modified foodstuff be passed to bacteria in the human gut, making these resistant to kanamycin and related antibiotics?

(2) Could the kan^R gene be passed to other organisms in the environment and would this result in damage to the ecosystem?

Neither question can be fully answered with our current knowledge. It can be argued that digestive processes would destroy all the kan^R genes in a genetically modified food before they could reach the bacterial flora of the gut, and that, even if a gene did avoid destruction, the chances of it being transferred to a bacterium would be very small. Nevertheless, the risk factor is not zero. Similarly, although experiments suggest that growth of genetically modified plants would have a negligible effect on the environment, as kan^R genes are already common in natural ecosystems, the future occurrence of some unforeseen and damaging event cannot be considered an absolute impossibility.

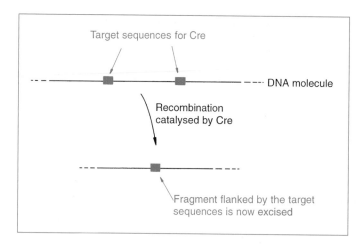

Figure 15.10 DNA excision by the Cre recombinase enzyme.

The fears surrounding the use of *kan^R* and other marker genes have prompted biotechnologists to devise ways of removing these genes from plant DNA after the transformation event has been verified. One of the strategies makes use of an enzyme from bacteriophage P1, called Cre, which catalyses a recombination event that excises DNA fragments flanked by specific 34 bp recognition sequences (Figure 15.10). To use this system the plant is transformed with two cloning vectors, the first carrying the gene being added to the plant along with its *kan^R* selectable marker gene surrounded by the Cre target sequences, and the second carrying the Cre gene. After transformation, expression of the Cre gene results in excision of the *kan^R* gene from the plant DNA.

What if the Cre gene is itself hazardous in some way? This is immaterial as the two vectors used in the transformation would probably integrate their DNA fragments into different chromosomes, so random segregation during sexual reproduction would result in first generation plants that contained one integrated fragment but not the other. A plant that contains neither the Cre gene nor the *kan^R* selectable marker, but does contain the important gene that we wished to add to the plant's genome, can therefore be obtained.

15.3.2

The possibility of harmful effects on the environment

A second area of concern regarding genetically modified plants is that their new gene combinations might harm the environment in some way. Particular problems have been highlighted with plants engineered to be resistant to virus infection. One approach here is to transform the plant with genes coding for coat proteins of a pathogenic virus. Expression of these genes does not result in disease symptoms but does provide the plant with a degree of protection against infection by the intact virus. One fear is that a plant that is synthesizing coat proteins for a pathogenic virus might be attacked by a second type of virus, the replication of which might lead to hybrid progeny containing genomes of the infecting virus packaged into coat proteins synthesized by the plant. These hybrids might have unexpected and dangerous properties, for example the new coat proteins might extend the host range of the second virus, enabling it to infect new plants that are normally resistant to the resulting disease.

There is also the question of whether a new gene introduced into a genetically modified plant could 'escape' and colonize wild plants and, if so, what impact this would have on the ecosystem. Research with natural and genetically modified plants is gradually accumulating the data from which these questions will be answered. But it is important to realize that determining the effects that genetically modified plants can have on the environment will be possible only if suitable experiments are carried out, in model systems, before large scale release of the plants is allowed. These experiments will be incomplete if they do not involve trials of genetically modified plants grown under carefully controlled conditions in the natural environment. If it is not possible to conduct these trials properly then harmful effects might be missed.

Further reading

Bachem, C.W.B., Speckmann, G.J., van der Linde, P.C.G. *et al.* (1994) Antisense expression of polyphenol oxidase genes inhibits enzymatic browning in potato tubers. *Biotechnology*, **12**, 1101–5.

Courtney-Gutterson, N., Napoli, C., Lemieux, C., Morgan, A., Firoozabady, E. & Robinson, K.E.P. (1994) Modification of flower colour in florist's chrysanthemum: production of a white-flowering variety through molecular genetics. *Biotechnology*, **12**, 268–71. [Another example of the use of antisense RNA.]

Feitelson, J.S., Payne, J. & Kim, L. (1992) *Bacillus thuringiensis*: insects and beyond. *Biotechnology*, **10**, 271–5. [Details of δ-endotoxins and their potential as conventional insecticides and in genetic engineering.]

Fischhoff, D.A., Bowdish, K.S., Perlak, F.J. *et al.* (1987) Insect-tolerant transgenic tomato plants. *Biotechnology*, **5**, 807–13. [The first transfer of a *B. thuringiensis* δ-endotoxin gene to a plant.]

Flavell, R.B., Dart, E., Fuchs, R.L. & Fraley, R.T. (1992) Selectable marker genes: safe for plants? *Biotechnology*, **10**, 141–4. [Describes the possible hazards posed by *kan*[R] and other marker genes in engineered plants.]

Koziel, M.G., Beland, G.L., Bowman, C. *et al.* (1993) Field performance of elite transgenic maize plants expressing an insecticidal protein derived from *Bacillus thuringiensis*. *Biotechnology*, **11**, 194–200.

Shade, R.E., Schroeder, H.E., Pueyo, J.J. *et al.* (1994) Transgenic pea seeds expressing the α-amylase inhibitor of the common bean are resistant to bruchid beetles. *Biotechnology*, **12**, 793–6. [A second approach to insect-resistant plants.]

Smith, C.J.S., Watson, C.F., Ray, J. *et al.* (1988) Antisense RNA inhibition of polygalacturonase gene expression in transgenic tomatoes. *Nature*, **334**, 724–6.

Tepfer, M. (1993) Viral genes and transgenic plants: what are the potential environmental risks? *Biotechnology*, **11**, 1125–32. [An excellent exploration of this difficult topic.]

Truve, E., Aaspõllu, A., Honkanen, J. *et al.* (1993) Transgenic potato plants expressing mammalian 2′–5′ oligoadenylate synthetase are protected from potato virus X infection under field conditions. *Biotechnology*, **11**, 1048–52. [Engineering virus resistance by gene addition.]

Yoder, J.I. & Goldsbrough, A.P. (1994) Transformation systems for generating marker-free transgenic plants. *Biotechnology*, **12**, 263–7. [Includes a description of the Cre excision system.]

Chapter 16

Gene Cloning and DNA Analysis in Forensic Science

Forensic science is the final area of biotechnology that we will consider. Hardly a week goes by without a report in the national press of another high profile crime that has been solved thanks to DNA analysis. The applications of molecular biology in forensics centre largely on the ability of DNA analysis to identify an individual from hairs, bloodstains and other items recovered from the crime scene. In the popular media, these techniques are called **genetic fingerprinting**, though the more accurate term for the procedures used today is **DNA profiling**. We begin this chapter by examining the methods used in genetic fingerprinting and DNA profiling, including their use both in identification of individuals and in establishing if individuals are members of a single family. This will lead us into an exploration of the ways in which genetic techniques are being used in areas of forensics other than police work, notably in archaeology.

16.1   DNA analysis in the identification of crime suspects

It is probably impossible for a person to commit a crime without leaving behind a trace of his or her DNA. Hairs, spots of blood and even conventional fingerprints contain traces of DNA, enough to be studied by the polymerase chain reaction (PCR). The analysis does not have to be done immediately, and in recent years a number of past crimes have been solved and the criminal brought to justice because of DNA testing that has been carried out on archived material. So how do these powerful methods work?

The basis to genetic fingerprinting and DNA profiling is that identical twins are the only individuals who have identical copies of the human genome. Of course, the human genome is more or less the same in everybody – the same genes will be in the same order with the same stretches of intergenic DNA between them. But the human genome, as well as those of other organisms, contains many **polymorphisms**, positions where the nucleotide sequence is not the same in every member of the population. We have already encountered the most important of these polymorphic sites earlier, because these variable sequences are the same ones that are used as DNA markers in genome mapping (p. 255). They include restriction fragment length polymorphisms (RFLPs), short tandem repeats (STRs) and single nucleotide polymorphisms (SNPs). All three can occur within genes as well as in intergenic regions, and altogether there are several million of these polymorphic sites in the human genome, with SNPs being the most common.

16.1.1   Genetic fingerprinting by hybridization probing

The first method for using DNA analysis to identify individuals was developed in the mid-1980s by Sir Alec Jeffreys of Leicester University. This technique was not based on any of the types of polymorphic site listed above, but on a different kind of variation in the human genome called a **hypervariable dispersed repetitive sequence**. As the name indicates, this is a repeated sequence that occurs at various places ('dispersed') in the human genome. The key feature of these sequences is that their genomic positions are variable: they are located at different positions in the genomes of different people (Figure 16.1(a)).

The particular repeat that was initially used in genetic fingerprinting contains the sequence GGGCAGGANG (where N is any nucleotide). To prepare a fingerprint a sample of DNA is digested with a restriction endonuclease and the fragments separated by agarose gel electrophoresis and a Southern blot prepared (p. 198). Hybridization to the blot of a labelled probe containing the repeat sequence reveals a series of bands, each one representing a restriction fragment that contains the repeat (Figure 16.1(b)). Because the insertion sites of the repeat sequence are variable, the same procedure carried out with a DNA sample from a second person will give a different pattern of bands. These are the genetic fingerprints for those individuals.

16.1.2   DNA profiling by PCR of short tandem repeats

Strictly speaking, genetic fingerprinting refers only to hybridization analysis of dispersed repeat sequences. This technique has been valuable in forensic work but suffers from three limitations:

(1) A relatively large amount of DNA is needed because the technique depends on hybridization analysis. Fingerprinting cannot be used with the minute amounts of DNA in hair and bloodstains.

(2) Interpretation of the fingerprint can be difficult because of variations in the intensities of the hybridization signals. In a court of law, minor

Figure 16.1 Genetic fingerprinting. (a) The positions of polymorphic repeats, such as hypervariable dispersed repetitive sequences, in the genomes of two individuals. In the chromosome segments shown, the second person has an additional repeat sequence. (b) An autoradiograph showing the genetic fingerprints of two individuals.

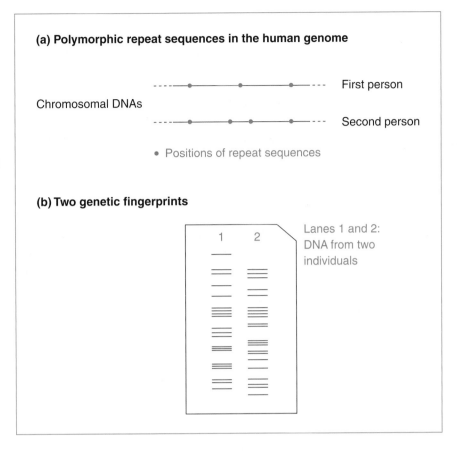

(a) Polymorphic repeat sequences in the human genome

Chromosomal DNAs

First person

Second person

• Positions of repeat sequences

(b) Two genetic fingerprints

Lanes 1 and 2: DNA from two individuals

differences in band intensity between a test fingerprint and that of a suspect can be sufficient for the suspect to be acquitted.

(3)   Although the insertion sites of the repeat sequences are hypervariable, there is a limit to this variability and therefore a small chance that two unrelated individuals could have the same, or at least very similar, fingerprints. Again, this consideration can lead to acquittal when a case is brought to court.

The more powerful technique of DNA profiling avoids these problems. Profiling makes use of the polymorphic sequences called STRs. As described on p. 255, an STR is a short sequence, one to 13 nucleotides in length, that is repeated several times in a tandem array. In the human genome, the most common type of STR is the dinucleotide repeat $[CA]_n$, where 'n', the number of repeats, is usually between 5 and 20 (Figure 16.2(a)).

The number of repeats in a particular STR is variable because repeats can be added or, less frequently, removed by errors that occur during DNA replication. In the population as a whole, there might be as many as ten different versions of a particular STR, each of the alleles characterized by a different number of repeats. In DNA profiling the alleles of a selected number

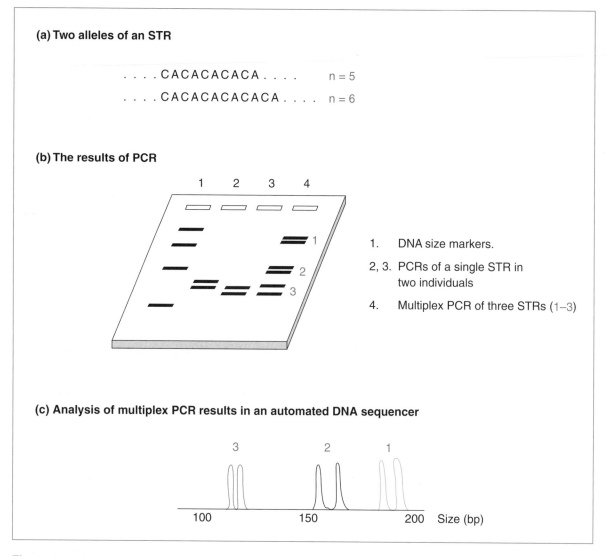

(a) Two alleles of an STR

. . . . CACACACACA . . . . $n = 5$

. . . . CACACACACACA . . . . $n = 6$

(b) The results of PCR

1   2   3   4

1.   DNA size markers.

2, 3.   PCRs of a single STR in two individuals

4.   Multiplex PCR of three STRs (1–3)

(c) Analysis of multiplex PCR results in an automated DNA sequencer

3   2   1

100   150   200   Size (bp)

Figure 16.2 DNA profiling. (a) DNA profiling makes use of STRs which have variable repeat units. (b) A gel obtained after DNA profiling. In lanes 2 and 3 the same STR has been examined in two individuals. These two people have different profiles, but have a band in common. Lane 4 shows the result of a multiplex PCR in which three STRs have been typed in a single PCR. (c) An automated DNA sequencer can be used to determine the sizes of the PCR products.

of different STRs are determined. This can be achieved quickly and with very small amounts of DNA by PCRs with primers that anneal to the DNA sequences either side of a repeat (Figure 12.11). After the PCR, the products are examined by agarose gel electrophoresis, with the size of the band or bands indicating the allele or alleles present in the DNA sample that has been tested (Figure 16.2(b)). Two alleles of an STR can be present in a single DNA sample because there are two copies of each STR, one on the chromosome inherited from the mother and one on the chromosome from the father.

Because PCR is used, DNA profiling is very sensitive and enables results to be obtained with hairs and other specimens that contain just trace amounts of DNA. The results are unambiguous, and a match between DNA profiles is usually accepted as evidence in a trial. Importantly, DNA profiling, when directed at STRs with large numbers of alleles, gives a very high statistical probability that a match between a test profile and that of a suspect is significant and not due to a chance similarity between two different people. The necessary degree of certainty can be achieved by analysis of a panel of nine STRs, which can be typed in a single **multiplex PCR**, in which a series of primer pairs are used in a single reaction. The results can be interpreted because the PCRs are designed so that the products obtained from each STR have different sizes and so appear at different positions on the agarose gel (Figure 16.2(b)). Alternatively, if the primers are labelled with different fluorochromes, the results can be visualized by running the products in an automated DNA sequencer (Figure 16.2(c)).

16.2 Studying kinship by DNA profiling

As well as identification of criminals, DNA profiling can also be used to infer if two or more individuals are members of the same family. This type of study is called **kinship analysis** and its main day-to-day application is in paternity testing.

16.2.1 Related individuals have similar DNA profiles

Your DNA profile, like all other aspects of your genome, is inherited partly from your mother and partly from your father. Relationships within a family therefore become apparant when the alleles of a particular STR are marked on the family pedigree (Figure 16.3). In this example, we see that three of the four children have inherited the 12-repeat allele from the father. This observation in itself is not sufficient to deduce that these three children are siblings, though the statistical chance would be quite high if the 12-repeat allele was uncommon in the population as whole. To increase the degree of certainty, more STRs would have to be typed but, as with identification of individuals, the analysis need not be endless, because a comparison of nine STRs gives an acceptable probability that relationships that are observed are real.

16.2.2 DNA profiling and the remains of the Romanovs

An interesting example of the use of DNA profiling in a kinship study is provided by work carried out during the 1990s on the bones of the Romanovs, the last members of the Russian ruling family. Tsar Nicholas II was deposed

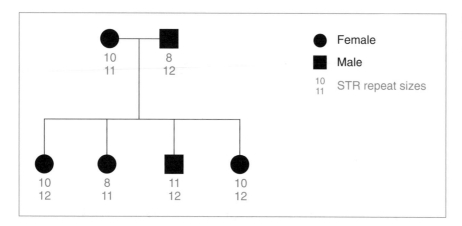

Figure 16.3 Inheritance of STR alleles within a family.

at the time of the Russian Revolution and he and his wife, the Tsarina Alexandra, and their five children were imprisoned. Then, in 1917, all seven were killed along with their doctor and various servants. In 1991, after the fall of communism, the bodies were recovered from their roadside grave.

STR analysis of the Romanov bones

The bodies that were recovered were little more than a collection of bones, those of the adults and children intermingled with no indications as to which belonged to the Romanovs and which to their doctor and servant. However, the only juvenile bones among the collection should have belonged to the children of the Tsar and Tsarina. This means that the bones of the Tsar and Tsarina could be identified by establishing which of the adults could be the parents of the children.

DNA was extracted from bones from each individual and five STRs typed by PCR. In fact, just two of these STRs provided sufficient information for the male and female parents of the children to be identified unambiguously (Figure 16.4). But were these indeed the bones of the Romanovs or could they be the remains some other unfortunate group of people? To address this problem the DNA from the bones was compared with DNA samples from living relatives of the Romanovs. This work included studies of **mitochondrial DNA**, the small 16 kb circles of DNA contained in the energy-generating mitochondria of cells. Mitochondrial DNA contains polymorphisms that can be used to infer relationships between individuals, but the degree of variability is not as great as displayed by STRs, so mitochondrial DNA is rarely used for kinship studies among closely related individuals such as those of a single family group. But mitochondrial DNA has the important property of being inherited solely through the female line, the father's mitochondrial DNA being lost during fertilization and not contributing to the son or daughter's DNA content. This maternal inheritance pattern makes it easier to distinguish relationships when the individuals being compared are more

Figure 16.4 Short tandem repeat analysis of the Romanov bones. (a) The Romanov family tree. (b) The results of STR analysis. *THO1* and *VWA/31* are the names of two STR loci. The numbers in the columns (8, 10; etc) are the repeat numbers for the alleles typed in each individual. The *THO1* data show that female adult 2 cannot be the mother of the children because she only possesses allele 6, which none of the children have. Female adult 1, however, has allele 8, which all three children have, and so is identified as the Tsarina. The *THO1* data exclude male adult 4 as a possible father of the children, but do not allow the other three male adults to be distinguished – all could be the father of at least two of the children. However, the *VWA/31* results exclude male adults 1 and 2, so male adult 3 is identified as the Tsar.

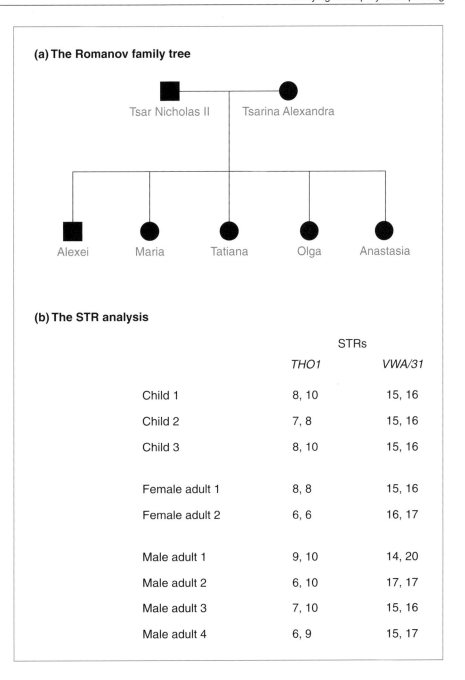

(a) The Romanov family tree

Tsar Nicholas II   Tsarina Alexandra

Alexei   Maria   Tatiana   Olga   Anastasia

(b) The STR analysis

	STRs	
	THO1	*VWA/31*
Child 1	8, 10	15, 16
Child 2	7, 8	15, 16
Child 3	8, 10	15, 16
Female adult 1	8, 8	15, 16
Female adult 2	6, 6	16, 17
Male adult 1	9, 10	14, 20
Male adult 2	6, 10	17, 17
Male adult 3	7, 10	15, 16
Male adult 4	6, 9	15, 17

distantly related, as was the case with the living relatives of the Romanovs. These mitochondrial DNA studies showed that the bones were indeed those of Tsar Nicholas, Tsarina Alexandra, and three of their daughters.

The missing children

Only three children were found in the Romanovs' grave. Alexei, the only boy, and one of the four girls were missing. During the middle decades of

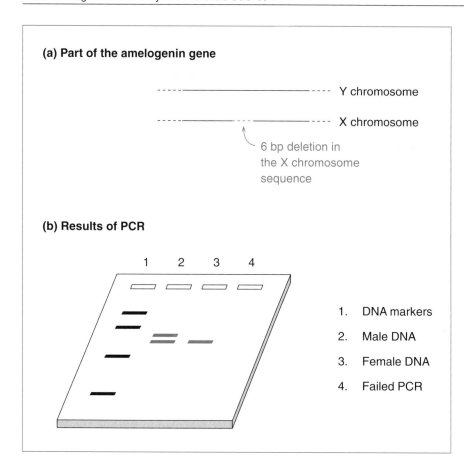

(a) Part of the amelogenin gene

Y chromosome

X chromosome

6 bp deletion in
the X chromosome
sequence

(b) Results of PCR

1 2 3 4

1. DNA markers

2. Male DNA

3. Female DNA

4. Failed PCR

Figure 16.6 Sex identification by PCR of part of the amelogenin gene. (a) An indel in the amelogenin gene. (b) The results of PCRs spanning the indel. Male DNA gives two PCR products, of 106 and 112 bp in the standard system used in forensics and biomolecular archaeology. Female DNA gives just the smaller product. A failed PCR gives no products and so is clearly distinguishable from the two types of positive result.

(Figure 16.6(a)). If the primers for a PCR anneal either side of an indel, the products obtained from the X and Y chromosomes would have different sizes. Female DNA would give a single band when the products are examined, because females only have the X chromosome, whereas males would give two bands, one from the X chromosome and one from the Y (Figure 16.6(b)). If the sample contains no DNA or the PCR fails for some other reason, no band will be obtained: there is no confusion between the failure and a male or female result.

The development of the amelogenin system for sex identification is having an important impact in archaeology. No longer is it necessary to assign sex to buried bones on the basis of vague differences in the structures of the bones. The greater confidence that DNA-based sex testing allows is resulting in some unexpected discoveries. In particular, archaeologists are now reviewing their preconceptions about the meaning of the objects buried in a grave along with the body. It was thought that if a body was accompanied by a sword then it must be male, or if the grave contained beads then the body was female. DNA testing has shown that these stereotypes are not always correct and that archaeologists must take a broader view of the link between grave goods and sex. Of all the outcomes of the recombinant DNA revolution, this is probably the one that was least expected when gene cloning was invented back in the 1970s.

Further reading

Brown, K.A. (1998) Gender and sex: what can ancient DNA tell us? *Ancient Biomolecules*, **2**, 3–15.

Gill, P., Ivanov, P.L., Kimpton, C. *et al.* (1994) Identification of the remains of the Romanov family by DNA analysis. *Nature Genetics*, **6**, 130–5.

Jeffreys, A.J., Wilson, V. & Thein, L.S. (1985) Individual-specific fingerprints of human DNA. *Nature*, **314**, 67–73. Genetic fingerprinting by analysis of minisatellites.

Krawczak, M. & Schmidtke, J. (1998) *DNA Fingerprinting*, 2nd edn. BIOS Scientific Publishers, Oxford.

Nakahori, Y., Hamano, K., Iwaya, M. & Nakagome, Y. (1991) Sex identification by polymerase chain reaction using X–Y homologous primer. *American Journal of Medical Genetics*, **39**, 472–3. The amelogenin method.

Biolistics   A means of introducing DNA into cells that involves bombardment with high velocity microprojectiles coated with DNA.

Biological containment   One of the precautionary measures taken to prevent the replication of recombinant DNA molecules in microorganisms in the natural environment. Biological containment involves the use of vectors and host organisms that have been modified so that they will not survive outside of the laboratory.

Biotechnology   The use of biological processes in industry and technology.

Biotin   A molecule that can be incorporated into dUTP and used as a non-radioactive label for a DNA probe.

BLAST   An algorithm frequently used in homology searching.

Blunt end or **flush end**   An end of a DNA molecule at which both strands terminate at the same nucleotide position with no single-stranded extension.

Broad host range plasmid   A plasmid that can replicate in a variety of host species.

Broth culture   Growth of microorganisms in a liquid medium.

Buoyant density   The density possessed by a molecule or particle when suspended in an aqueous salt or sugar solution.

Candidate gene   A gene, identified by positional cloning, that might be a disease-causing gene.

Capsid   The protein coat that encloses the DNA or RNA molecule of a bacteriophage or virus.

Cassette   A DNA sequence consisting of promoter–ribosome binding site–unique restriction site–terminator (or for a eukaryotic host, promoter–unique restriction site–polyadenylation sequence) carried by certain types of expression vector. A foreign gene inserted into the unique restriction site is placed under control of the expression signals.

Cauliflower mosaic virus (CaMV)   The best studied of the caulimoviruses, used in the past as a cloning vector for some species of higher plant. Cauliflower mosaic virus is the source of strong promoters used in other types of plant cloning vector.

Caulimoviruses   One of the two groups of DNA viruses to infect plants, the members of which have potential as cloning vectors for some species of higher plant.

Cell extract   A preparation consisting of a large number of broken cells and their released contents.

Cell-free translation system   A cell extract containing all the components required for protein synthesis (i.e. ribosomal subunits, tRNAs, amino acids, enzymes and cofactors) and able to translate added mRNA molecules.

Chimera   (1) A recombinant DNA molecule made up of DNA fragments from more than one organism, named after the mythological beast.

(*Continued.*) (2) The initial product of cloning using embryonic stem cells: an animal made up of a mixture of cells with different genotypes.

Chromosome One of the DNA–protein structures that contains part of the nuclear genome of a eukaryote. Less accurately, the DNA molecule(s) that contains a prokaryotic genome.

Chromosome walking A technique that can be used to construct a clone contig by identifying overlapping fragments of cloned DNA.

Cleared lysate A cell extract that has been centrifuged to remove cell debris, subcellular particles and possibly chromosomal DNA.

Clone A population of identical cells, generally those containing identical recombinant DNA molecules.

Clone contig approach A genome sequencing strategy in which the molecules to be sequenced are broken into manageable segments, each a few hundred kilobases or a few megabases in length, which are sequenced individually.

Clone fingerprinting Any one of a variety of techniques that compares cloned DNA fragments in order to identify ones that overlap.

Codon bias The fact that not all codons are used equally frequently in the genes of a particular organism.

Combinatorial screening A technique that reduces the number of PCRs or other analyses that must be performed by combining samples in an ordered fashion, so that a sample giving a particular result can be identified even though that sample is not individually examined.

Compatibility The ability of two different types of plasmid to coexist in the same cell.

Competent A culture of bacteria that has been treated to enhance their ability to take up DNA molecules.

Complementary Two polynucleotides that can base pair to form a double-stranded molecule.

Complementary DNA (cDNA) cloning A cloning technique involving conversion of purified mRNA to DNA before insertion into a vector.

Conformation The spatial organization of a molecule. Linear and circular are two possible conformations of a polynucleotide.

Conjugation Physical contact between two bacteria, usually associated with transfer of DNA from one cell to the other.

Consensus sequence A nucleotide sequence used to describe a large number of related though non-identical sequences. Each position of the consensus sequence represents the nucleotide most often found at that position in the real sequences.

Contig A contiguous segment of DNA sequence obtained as part of a genome sequencing project.

Continuous culture The culture of microorganisms in liquid medium under controlled conditions, with additions to and removals from the medium over a lengthy period of time.

Contour clamped homogeneous electric fields (CHEF)   An electrophoresis technique for the separation of large DNA molecules.

Copy number   The number of molecules of a plasmid contained in a single cell.

Cosmid   A cloning vector consisting of the λ *cos* site inserted into a plasmid, used to clone DNA fragments up to 40 kb in size.

***cos* site**   One of the cohesive, single-stranded extensions present at the ends of the DNA molecules of certain strains of λ phage.

Covalently closed-circular DNA (cccDNA)   A completely double-stranded circular DNA molecule, with no nicks or discontinuities, usually with a supercoiled conformation.

CpG island   A GC-rich DNA region located upstream of approximately 56% of the genes in the human genome.

Defined medium   A bacterial growth medium in which all the components are known.

Deletion analysis   The identification of control sequences for a gene by determining the effects on gene expression of specific deletions in the upstream region.

Denaturation   Of nucleic acid molecules: breakdown by chemical or physical means of the hydrogen bonds involved in base pairing.

Density gradient centrifugation   Separation of molecules and particles on the basis of buoyant density, by centrifugation in a concentrated sucrose or caesium chloride solution.

Deoxyribonuclease   An enzyme that degrades DNA.

Dideoxynucleotide   A modified nucleotide that lacks the 3′ hydroxyl group and so prevents further chain elongation when incorporated into a growing polynucleotide.

Direct gene transfer   A cloning process that involves transfer of a gene into a chromosome without the use of a cloning vector able to replicate in the host organism.

Disarmed plasmid   A Ti plasmid that has had some or all of the T-DNA genes removed, so it is no longer able to promote cancerous growth of plant cells.

DNA chip   A wafer of silicon carrying a high density array of oligonucleotides used in transcriptome and other studies.

DNA marker   A DNA sequence that exists as two or more alleles and which can therefore be used in genetic mapping.

DNA polymerase   An enzyme that synthesizes DNA on a DNA or RNA template.

DNA profiling   A PCR technique that determines the alleles present at different STR loci within a genome in order to use DNA information to identify individuals.

DNA sequencing   Determination of the order of nucleotides in a DNA molecule.

Double digestion Cleavage of a DNA molecule with two different restriction endonucleases, either concurrently or consecutively.

Electrophoresis Separation of molecules on the basis of their charge-to-mass ratio.

Electroporation A method for increasing DNA uptake by protoplasts through prior exposure to a high voltage, which results in the temporary formation of small pores in the cell membrane.

Embryonic stem (ES) cell A totipotent cell from the embryo of a mouse or other organism, used in construction of a transgenic animal such as a knockout mouse.

End filling Conversion of a sticky end to a blunt end by enzymatic synthesis of the complement to the single-stranded extension.

Endonuclease An enzyme that breaks phosphodiester bonds within a nucleic acid molecule.

Episome A plasmid capable of integration into the host cell's chromosome.

Ethanol precipitation Precipitation of nucleic acid molecules by ethanol plus salt, used primarily as a means of concentrating DNA.

Ethidium bromide A fluorescent chemical that intercalates between base pairs in a double-stranded DNA molecule, used in the detection of DNA.

Exonuclease An enzyme that sequentially removes nucleotides from the ends of a nucleic acid molecule.

Expressed sequence tag (EST) A partial or complete cDNA sequence.

Expression vector A cloning vector designed so that a foreign gene inserted into the vector is expressed in the host organism.

Fermenter A vessel used for the large scale culture of microorganisms.

Field inversion gel electrophoresis (FIGE) An electrophoresis technique for the separation of large DNA molecules.

Fluorescence *in situ* hybridization (FISH) A hybridization technique that uses fluorochromes of different colours to enable two or more genes to be located within a chromosome preparation in a single *in situ* experiment.

Footprinting The identification of a protein binding site on a DNA molecule by determining which phosphodiester bonds are protected from cleavage by DNase I.

Functional genomics Studies aimed at identifying all the genes in a genome and determining their functions.

Gel electrophoresis Electrophoresis performed in a gel matrix so that molecules of similar electric charge can be separated on the basis of size.

Gel retardation A technique that identifies a DNA fragment that has a bound protein by virtue of its decreased mobility during gel electrophoresis.

Geminivirus One of the two groups of DNA viruses that infect plants, the members of which have potential as cloning vectors for some species of higher plants.

Gene   A segment of DNA that codes for an RNA and/or polypeptide molecule.

Gene addition   A genetic engineering strategy that involves the introduction of a new gene or group of genes into an organism.

Gene cloning   Insertion of a fragment of DNA, carrying a gene, into a cloning vector, and subsequent propagation of the recombinant DNA molecule in a host organism. Also used to describe those techniques that achieve the same result without the use of a cloning vector (e.g. direct gene transfer).

Gene knockout   A technique that results in inactivation of a gene, as a means of determining the function of that gene.

Gene mapping   Determination of the relative positions of different genes on a DNA molecule.

Gene subtraction   A genetic engineering strategy that involves the inactivation of one or more of an organism's genes.

Gene therapy   A clinical procedure in which a gene or other DNA sequence is used to treat a disease.

Genetic engineering   The use of experimental techniques to produce DNA molecules containing new genes or new combinations of genes.

Genetic fingerprinting   A hybridization technique that determines the genomic distribution of a hypervariable dispersed repetitive sequence and results in a banding pattern that is specific for each individual.

Genetic map   A genome map that has been obtained by analysing the results of genetic crosses.

Genetics   The branch of biology devoted to the study of genes.

Genome   The complete set of genes of an organism.

Genomic library   A collection of clones sufficient in number to include all the genes of a particular organism.

Genomics   The study of a genome, in particular the complete sequencing of a genome.

GM (genetically modified) crop   A crop plant that has been engineered by gene addition or gene subtraction.

Harvesting   The removal of microorganisms from a culture, usually by centrifugation.

Helper phage   A phage that is introduced into a host cell in conjunction with a related cloning vector, in order to provide enzymes required for replication of the cloning vector.

Heterologous probing   The use of a labelled nucleic acid molecule to identify related molecules by hybridization probing.

Homology   Refers to two genes from different organisms that have evolved from the same ancestral gene. Two homologous genes are usually sufficiently similar in sequence for one to be used as a hybridization probe for the other.

Homology search A technique in which genes with sequences similar to that of an unknown gene are sought, in order to confirm a gene identification or to understand the function of the unknown gene.

Homopolymer tailing The attachment of a sequence of identical nucleotides (e.g. AAAAA) to the end of a nucleic acid molecule, usually referring to the synthesis of single-stranded homopolymer extensions on the ends of a double-stranded DNA molecule.

Horseradish peroxidase An enzyme that can be complexed to DNA and which is used in a non-radioactive procedure for DNA labelling.

Host-controlled restriction A mechanism by which some bacteria prevent phage attack through the synthesis of a restriction endonuclease that cleaves the non-bacterial DNA.

Hybrid-arrest translation (HART) A method used to identify the polypeptide coded by a cloned gene.

Hybridization probe A labelled nucleic acid molecule that can be used to identify complementary or homologous molecules through the formation of stable base-paired hybrids.

Hybrid-release translation (HRT) A method used to identify the polypeptide coded by a cloned gene.

Hypervariable dispersed repetitive sequence The type of human repetitive DNA sequence used in genetic fingerprinting.

Immunological screening The use of an antibody to detect a polypeptide synthesized by a cloned gene.

Inclusion body A crystalline or paracrystalline deposit within a cell, often containing substantial quantities of insoluble protein.

Incompatibility group Comprises a number of different types of plasmid, often related to each other, that are unable to coexist in the same cell.

Indel A position where a DNA sequence has been inserted into or deleted from a genome, so-called because it is impossible from comparison of two sequences to determine which alternative has occurred – insertion into one genome or deletion from the other.

Induction (1) Of a gene: the switching on of the expression of a gene or group of genes in response to a chemical or other stimulus.

(2) Of λ phage: the excision of the integrated form of λ and accompanying switch to the lytic mode of infection, in response to a chemical or other stimulus.

Insertional inactivation A cloning strategy whereby insertion of a new piece of DNA into a vector inactivates a gene carried by the vector.

Insertion vector A λ vector constructed by deleting a segment of non-essential DNA.

***In situ* hybridization** A technique for gene mapping involving hybridization of a labelled sample of a cloned gene to a large DNA molecule, usually a chromosome.

Interspersed repeat element PCR (IRE–PCR) A clone fingerprinting technique that uses PCR to detect the relative positions of repeated sequences in cloned DNA fragments.

***In vitro* mutagenesis** Any one of several techniques used to produce a specified mutation at a predetermined position in a DNA molecule.

***In vitro* packaging** Synthesis of infective λ particles from a preparation of λ capsid proteins and a catenane of DNA molecules separated by *cos* sites.

Kinship analysis An examination of DNA profiles or other information to determine if two individuals are related.

Klenow fragment (of DNA polymerase I) A DNA polymerase enzyme, obtained by chemical modification of *E. coli* DNA polymerase I, used primarily in chain termination DNA sequencing.

Knockout mouse A mouse that has been engineered so that it carries an inactivated gene.

Labelling The incorporation of a marker nucleotide into a nucleic acid molecule. The marker is often but not always a radioactive or fluorescent label.

Lac selection A means of identifying recombinant bacteria containing vectors that carry the *lacZ′* gene. The bacteria are plated on a medium that contains an analogue of lactose that gives a blue colour in the presence of β-galactosidase activity.

Lambda (λ) A bacteriophage that infects *E. coli*, derivatives of which are used as cloning vectors.

Ligase (DNA ligase) An enzyme that, in the cell, repairs single-stranded discontinuities in double-stranded DNA molecules. Purified DNA ligase is used in gene cloning to join DNA molecules together.

Linkage analysis A technique for mapping the chromosomal position of a gene by comparing its inheritance pattern with that of genes and other loci whose map positions are already known.

Linker A synthetic, double-stranded oligonucleotide used to attach sticky ends to a blunt-ended molecule.

Lysogen A bacterium that harbours a prophage.

Lysogenic infection cycle The pattern of phage infection that involves integration of the phage DNA into the host chromosome.

Lysozyme An enzyme that weakens the cell walls of certain types of bacteria.

Lytic infection cycle The pattern of infection displayed by a phage that replicates and lyses the host cell immediately after the initial infection. Integration of the phage DNA molecule into the bacterial chromosome does not occur.

M13 A bacteriophage that infects *E. coli*, derivatives of which are used as cloning vectors.

Mapping reagent   A collection of DNA fragments spanning a chromosome or the entire genome and used in STS mapping.

Melting temperature (T_m)   The temperature at which a double-stranded DNA or DNA–RNA molecule denatures.

Messenger RNA (mRNA)   The transcript of a protein-coding gene.

Microarray   A set of genes or cDNAs immobilized on a glass slide and used in transcriptome studies.

Microinjection   A method of introducing new DNA into a cell by injecting it directly into the nucleus.

Microsatellite   A polymorphism comprising tandem copies of, usually, two-, three-, four- or five-nucleotide repeat units. Also called a short tandem repeat (STR).

Minimal medium   A defined medium that provides only the minimum number of different nutrients needed for growth of a particular bacterium.

Mitochondrial DNA   The DNA molecules present in the mitochondria of eukaryotes.

Modification interference assay   A technique that uses chemical modification to identify nucleotides involved in interactions with a DNA-binding protein.

Multicopy plasmid   A plasmid with a high copy number.

Multigene family   A number of identical or related genes present in the same organism, usually coding for a family of related polypeptides.

Multiplex PCR   A PCR carried out with more than one pair of primers and hence targeting two or more sites in the DNA being studied.

2 μm circle   A plasmid found in the yeast *Saccharomyces cerevisiae* and used as the basis for a series of cloning vectors.

Nick   A single-strand break, involving the absence of one or more nucleotides, in a double-stranded DNA molecule.

Nick translation   The repair of a nick with DNA polymerase I, usually to introduce labelled nucleotides into a DNA molecule.

Northern transfer   A technique for transferring bands of RNA from an agarose gel to a nitrocellulose or nylon membrane.

Nucleic acid hybridization   Formation of a double-stranded molecule by base pairing between complementary or homologous polynucleotides.

Oligonucleotide   A short, synthetic, single-stranded DNA molecule, such as one used as a primer in DNA sequencing or PCR.

Oligonucleotide-directed mutagenesis   An *in vitro* mutagenesis technique that involves the use of a synthetic oligonucleotide to introduce the predetermined nucleotide alteration into the gene to be mutated.

Open-circular DNA (ocDNA)   The non-supercoiled conformation taken up by a circular double-stranded DNA molecule when one or both polynucleotides carry nicks.

Open reading frame (ORF)   A series of codons that is or could be a gene.

Origin of replication The specific position on a DNA molecule where DNA replication begins.

Orphan An open reading frame thought to be a functional gene but to which no function has yet been assigned.

Orthogonal field alternation gel electrophoresis (OFAGE) A gel electrophoresis technique that employs a pulsed electric field to achieve separation of very large molecules of DNA.

P1 A bacteriophage that infects *E. coli*, derivatives of which are used as cloning vectors.

P1-derived artificial chromosome (PAC) A cloning vector based on the P1 bacteriophage, used for cloning relatively large fragments of DNA in *E. coli*.

Papillomaviruses A group of mammalian viruses, derivatives of which have been used as cloning vectors.

Partial digestion Treatment of a DNA molecule with a restriction endonuclease under such conditions that only a fraction of all the recognition sites are cleaved.

Pedigree analysis The use of a human family tree to analyse the inheritance of a genetic or DNA marker.

P element A transposon from *Drosophila melanogaster* used as the basis of a cloning vector for that organism.

Phage display A technique involving cloning in M13 that is used to identify proteins that interact with one another.

Phage display library A collection of M13 clones carrying different DNA fragments, used in phage display.

Phagemid A double-stranded plasmid vector that possesses an origin of replication from a filamentous phage and hence can be used to synthesize a single-stranded version of a cloned gene.

Pharming Genetic modification of a farm animal so that the animal synthesizes a recombinant pharmaceutical protein, often in its milk.

Physical map A genome map that has been obtained by direct examination of DNA molecules.

Pilus One of the structures present on the surface of a bacterium containing a conjugative plasmid, through which DNA is assumed to pass during conjugation.

Plaque A zone of clearing on a lawn of bacteria caused by lysis of the cells by infecting phage particles.

Plasmid A usually circular piece of DNA, primarily independent of the host chromosome, often found in bacteria and some other types of cells.

Plasmid amplification A method involving incubation with an inhibitor of protein synthesis aimed at increasing the copy number of certain types of plasmid in a bacterial culture.

Polyethylene glycol A polymeric compound used to precipitate macromolecules and molecular aggregates.

Polylinker   A synthetic double-stranded oligonucleotide carrying a number of restriction sites.

Polymerase chain reaction (PCR)   A technique that enables multiple copies of a DNA molecule to be generated by enzymatic amplification of a target DNA sequence.

Polymorphism   Refers to a locus that is present as a number of different alleles or other variations in the population as a whole.

Positional cloning   A procedure that uses information on the map position of a gene to obtain a clone of that gene.

Positional effect   Refers to the variations in expression levels observed for genes inserted at different positions in a genome.

Post-genomics   Studies aimed at identifying all the genes in a genome and determining their functions.

Primer   A short single-stranded oligonucleotide which, when attached by base pairing to a single-stranded template molecule, acts as the start point for complementary strand synthesis directed by a DNA polymerase enzyme.

Primer extension   A method of transcript analysis in which the 5′ end of an RNA is mapped by annealing and extending an oligonucleotide primer.

Promoter   The nucleotide sequence, upstream of a gene, that acts as a signal for RNA polymerase binding.

Prophage   The integrated form of the DNA molecule of a lysogenic phage.

Protease   An enzyme that degrades protein.

Protein A   A protein from the bacterium *Staphylococcus aureus* that binds specifically to immunoglobulin G (i.e. antibody) molecules.

Protein engineering   A collection of techniques, including but not exclusively gene mutagenesis, that result in directed alterations being made to protein molecules, often to improve the properties of enzymes used in industrial processes.

Proteome   The entire protein content of a cell or tissue.

Protoplast   A cell from which the cell wall has been completely removed.

RACE (rapid amplification of cDNA ends)   A PCR-based technique for mapping the end of an RNA molecule.

Radiation hybrid   A collection of rodent cell lines that contain different fragments of the human genome, constructed by a technique involving irradiation and used as a mapping reagent in studies of the human genome.

Radioactive marker   A radioactive atom used in the detection of a larger molecule into which it has been incorporated.

Random priming   A method for DNA labelling that utilizes random DNA hexamers, which anneal to single-stranded DNA and act as primers for complementary strand synthesis by a suitable enzyme.

Reading frame   One of the six overlapping sequences of triplet codons, three on each polynucleotide, contained in a segment of a DNA double helix.

Recombinant   A transformed cell that contains a recombinant DNA molecule.

Recombinant DNA molecule   A DNA molecule created in the test tube by ligating together pieces of DNA that are not normally contiguous.

Recombinant DNA technology   All of the techniques involved in the construction, study and use of recombinant DNA molecules.

Recombinant protein   A polypeptide that is synthesized in a recombinant cell as the result of expression of a cloned gene.

Recombination   The exchange of DNA sequences between different molecules, occurring either naturally or as a result of DNA manipulation.

Relaxed   Refers to the non-supercoiled conformation of open-circular DNA.

Repetitive DNA PCR   A clone fingerprinting technique that uses PCR to detect the relative positions of repeated sequences in cloned DNA fragments.

Replacement vector   A λ vector designed so that insertion of new DNA is by replacement of part of the non-essential region of the λ DNA molecule.

Replica plating   A technique whereby the colonies on an agar plate are transferred *en masse* to a new plate, on which the colonies grow in the same relative positions as before.

Replicative form (RF) of M13   The double-stranded form of the M13 DNA molecule found within infected *E. coli* cells.

Reporter gene   A gene whose phenotype can be assayed in a transformed organism, and which is used in, for example, deletion analyses of regulatory regions.

Repression   The switching off of expression of a gene or a group of genes in response to a chemical or other stimulus.

Restriction analysis   Determination of the number and sizes of the DNA fragments produced when a particular DNA molecule is cut with a particular restriction endonuclease.

Restriction endonuclease   An endonuclease that cuts DNA molecules only at a limited number of specific nucleotide sequences.

Restriction fragment length polymorphism (RFLP)   A mutation that results in alteration of a restriction site and hence a change in the pattern of fragments obtained when a DNA molecule is cut with a restriction endonuclease.

Restriction map   A map showing the positions of different restriction sites in a DNA molecule.

Retrovirus   A virus with an RNA genome, able to insert into a host chromosome, derivatives of which have been used to clone genes in mammalian cells.

Reverse transcriptase An RNA-dependent DNA polymerase, able to synthesize a complementary DNA molecule on a template of single-stranded RNA.

RFLP linkage analysis A technique that uses a closely linked RFLP as a marker for the presence of a particular allele in a DNA sample, often as a means of screening individuals for a defective gene responsible for a genetic disease.

Ribonuclease An enzyme that degrades RNA.

Ribosome binding site The short nucleotide sequence upstream of a gene, which after transcription forms the site on the mRNA molecule to which the ribosome binds.

Ri plasmid An *Agrobacterium rhizogenes* plasmid, similar to the Ti plasmid, used to clone genes in higher plants.

Reverse transcription–PCR (RT–PCR) A PCR technique in which the starting material is RNA. The first step in the procedure is conversion of the RNA to cDNA with reverse transcriptase.

Selectable marker A gene carried by a vector and conferring a recognizable characteristic on a cell containing the vector or a recombinant DNA molecule derived from the vector.

Selection A means of obtaining a clone containing a desired recombinant DNA molecule.

Sequence tagged site (STS) A DNA sequence whose position has been mapped in a genome.

Short tandem repeat (STR) A polymorphism comprising tandem copies of, usually, two-, three-, four- or five-nucleotide repeat units. Also called a microsatellite.

Shotgun approach A genome sequencing strategy in which the molecules to be sequenced are randomly broken into fragments which are then individually sequenced.

Shotgun cloning A cloning strategy that involves the insertion of random fragments of a large DNA molecule into a vector, resulting in a large number of different recombinant DNA molecules.

Shuttle vector A vector that can replicate in the cells of more than one organism (e.g. in *E. coli* and in yeast).

Simian virus 40 (SV40) A mammalian virus that has been used as the basis for a cloning vector.

Single nucleotide polymorphism (SNP) A point mutation that is carried by some individuals of a population.

Sonication A procedure that uses ultrasound to cause random breaks in DNA molecules.

Southern transfer A technique for transferring bands of DNA from an agarose gel to a nitrocellulose or nylon membrane.

Sphaeroplast A cell with a partially degraded cell wall.

Stem loop   A hairpin structure, consisting of a base-paired stem and a non-base-paired loop, that may form in a polynucleotide.

Sticky end   An end of a double-stranded DNA molecule where there is a single-stranded extension.

Strong promoter   An efficient promoter that can direct synthesis of RNA transcripts at a relatively fast rate.

Stuffer fragment   The part of a λ replacement vector that is removed during insertion of new DNA.

Supercoiled   The conformation of a covalently closed-circular DNA molecule, which is coiled by torsional strain into the shape taken by a wound-up elastic band.

***Taq* DNA polymerase**   The thermostable DNA polymerase that is used in PCR.

T-DNA   The portion of the Ti plasmid transferred to the plant DNA.

Temperature-sensitive mutation   A mutation that results in a gene product that is functional within a certain temperature range (e.g. at less than 30°C), but non-functional at different temperatures (e.g. above 30°C).

Template   A single-stranded polynucleotide (or region of a polynucleotide) that directs synthesis of a complementary polynucleotide.

Terminator   The short nucleotide sequence, downstream of a gene, that acts as a signal for termination of transcription.

5′ terminus   One of the two ends of a polynucleotide: that which carries the phosphate group attached to the 5′ position of the sugar.

3′ terminus   One of the two ends of a polynucleotide: that which carries the hydroxyl group attached to the 3′ position of the sugar.

Thermal cycle sequencing   A DNA sequencing method that uses PCR to generate chain-terminated polynucleotides.

Ti plasmid   The large plasmid found in those *Agrobacterium tumefaciens* cells able to direct crown gall formation on certain species of plants.

Total cell DNA   Consists of all the DNA present in a single cell or group of cells.

Transcript analysis   A type of experiment aimed at determining which portions of a DNA molecule are transcribed into RNA.

Transcriptome   The entire mRNA content of a cell or tissue.

Transfection   The introduction of purified virus DNA molecules into any living cell.

Transformation   The introduction of any DNA molecule into any living cell.

Transformation frequency   A measure of the proportion of cells in a population that are transformed in a single experiment.

Transgenic animal   An animal that possesses a cloned gene in all of its cells.

Transposon   A DNA sequence that is able to move from place to place within a genome.

Undefined medium   A growth medium in which not all the components have been identified.

UV absorbance spectrophotometry   A method for measuring the concentration of a compound by determining the amount of ultraviolet radiation absorbed by a sample.

Vector   A DNA molecule, capable of replication in a host organism, into which a gene is inserted to construct a recombinant DNA molecule.

Vehicle   Often used as a substitute for the word 'vector', emphasizing that the vector transports the inserted gene through the cloning experiment.

Virus chromosome   The DNA or RNA molecule(s) contained within a virus capsid and carrying the viral genes.

Watson–Crick rules   The base pairing rules that underlie gene structure and expression. A pairs with T, and G pairs with C.

Weak promoter   An inefficient promoter that directs synthesis of RNA transcripts at a relatively low rate.

Western transfer   A technique for transferring bands of protein from an electrophoresis gel to a membrane support.

Yeast artificial chromosome (YAC)   A cloning vector comprising the structural components of a yeast chromosome and able to clone very large pieces of DNA.

Yeast episomal plasmid (YEp)   A yeast vector carrying the 2 μm circle origin of replication.

Yeast integrative plasmid (YIp)   A yeast vector that relies on integration into the host chromosome for replication.

Yeast replicative plasmid (YRp)   A yeast vector that carries a chromosomal origin of replication.

Yeast two hybrid system   A technique involving cloning in *S. cerevisiae* that is used to identify proteins that interact with one another.

Zoo blot   A nitrocellulose or nylon membrane carrying immobilized DNAs of several species, used to determine if a gene from one species has homologues in the other species.

Index

Please note: The letters 'g', 'f' and 't' with a page number denote glossary, figure, and table, respectively.